PIPELINE
RISK
MANAGEMENT
MANUAL

PIPELINE
◆ RISK ◆
MANAGEMENT
MANUAL

W. Kent Muhlbauer

Gulf Publishing Company
Houston, London, Paris, Zurich, Tokyo

Pipeline Risk Management Manual

Gulf Publishing Company
Book Division
P.O. Box 2608, Houston, Texas 77252-2608

10 9 8 7 6 5 4 3 2 1

Library of Congress Cataloging-in-Publication Data

Muhlbauer, W. Kent.
    Pipeline risk management manual: a systematic
approach to loss prevention and risk assessment/W.
Kent Muhlbauer.
        p. cm.
    Includes bibliographical references and index.
    ISBN 0-88415-035-6
    1. Pipe lines—Safety measures. 2. Pipe
lines—Reliability.
    I. Title.
    TJ930.M84 1992   91-32612
    621.8'672—dc20   CIP

Printed on Acid Free Paper (∞)

Contents

Acknowledgments

The author wishes to express his gratitude to Jack C. W. Fraser, Charles O. Warner, and the other pipeline professionals who offered their ideas and expertise to this undertaking.

Preface

When a cross country pipeline is installed and operated, a hazard that would not otherwise be present has been introduced. Society generally accepts that the benefits of this hazard far outweigh the increased risk.

The title of this text implies that pipeline risk is something that should be managed. In order to manage something, we must thoroughly understand it.

While we will most likely never be able to accurately predict all pipeline failures, we can, however, pick out what we believe to be important factors that MAY be contributors to pipeline failures. Analyzing these factors and their interactions will give us insight into the relative potential for a failure.

Risk assessment doesn't have to be a calculation-intensive exercise in probabilistic theory. Such calculations are, after all, based upon probabilities that are of questionable benefit in rare-occurrence scenarios. A false precision is often assigned to numbers that are the result of detailed calculations. In reality, the margin of uncertainty is quite high because of the large number of assumptions required in such analyses.

The approach used in this book is to deviate from strict scientific procedure in building this risk model. In many situations, some risk aspects are based as much upon intuition as upon hard evidence. Rather than being seen as a detraction, the author believes that this approach strengthens the risk management process.

It will be shown in the following chapters that historical data, working experience, and common sense can be combined into a flexible risk management tool. A "tool" is the ultimate objective, here. The most sophisticated analysis that is studied once and then filed away is at best only a means to satisfy an intellectual curiosity. An easy-to-understand, easy-to-modify system of risk assessment can become a part of every-day design, business, and operations decisions.

Explanations are provided in this book to describe the reasoning behind the inclusion of the risk contributors and risk reducers. By way of these explanations, the non-pipeline professional can obtain a feel for many aspects of pipeline design, operation, and mainte-nance. It is hoped that such understanding will help in communica-tions between pipeline operators, regulators, insurers, and other people with interests in this industry.

W. Kent Muhlbauer

Introduction

A systematic approach to pipeline risk management is proposed here. This book is organized to serve as a guide for the person or persons who are actually performing pipeline risk assessments.

All of the risk evaluation items with their suggested scores are listed on p. xii in this Introduction. This list can be used as a checklist for the actual pipeline evaluations and for subsequent data retention or entry into a computer program.

Chapter One explains the reasoning behind the type of risk assessment proposed here. Other forms of risk assessment are listed as well as some concepts of risk assessment in general. Concepts of quality and cost management, as they relate to risk management, are also discussed.

Chapter Two provides the foundation for this risk assessment process. Basic assumptions of this model and the structure of the evaluation process are covered. Sectioning of the pipeline and classifying pipeline activities as attributes or preventions are addressed here.

Chapters Three through Six detail the pipeline activities and environmental characteristics that influence risk. Each chapter corresponds to an index which, in turn, corresponds to a historical cause of pipeline failures. These chapters show the suggested scoring for each item and the rationale behind including the item in the risk assessment.

Chapter Seven details the Leak Impact Factor which is the 'consequence' part of the risk equation. Product handled, population density, and other factors are combined here to assess the potential consequences of a pipeline failure.

Chapter Eight offers suggestions on handling the evaluation data. Managing the database, extracting information, and presenting the information in a meaningful format are all covered in this chapter.

The appendices are included as reference sections to allow the evaluator to have necessary equations and product information readily available.

Specific examples are included all through the text to assist the reader in understanding the risk scoring system. An overall example is also provided in Appendix E. This overall example ties the risk assessment together in one evaluation of a hypothetical pipeline section.

Risk Evaluation at a Glance

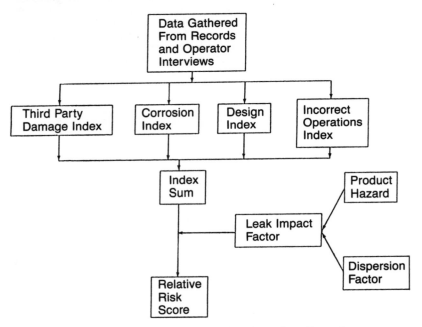

Figure I-1. Components of risk rating flowchart.

Relative Risk Rating = (Index Sum)/(Leak Impact Factor)

= [(Third Party Index)

+ (Corrosion Index)

+ (Design Index)

+ (Incorrect Operations Index)]

Leak Impact Factor

Third Party Index

A.	Minimum Depth of Cover	0–20 pts	20%
B.	Activity Level	0–20 pts	20%
C.	Aboveground Facilities	0–10 pts	10%
D.	One-Call System	0–15 pts	15%
E.	Public Education	0–15 pts	15%
F.	Right-of-Way Condition	0–5 pts	5%
G.	Patrol Frequency	0–15 pts	15%
		0–100 pts	100%

Corrosion Index

Corrosion Index =	(Atmospheric Corrosion)	20%
	+ (Internal Corrosion)	20%
	+ (Buried Metal Corrosion)	60%
		100%

A. Atmospheric Corrosion

1. Facilities	0–5 pts	
2. Atmosphere	0–10 pts	
3. Coating/Inspection	0–5 pts	
	0–20 pts	20%

B. Internal Corrosion

1. Product Corrosivity	0–10 pts	
2. Internal Protection	0–10 pts	
	0–20 pts	20%

C. Buried Metal Corrosion

1. Cathodic Protection	0–8 pts	
2. Coating Condition	0–10 pts	
3. Soil Corrosivity	0–4 pts	
4. Age of System	0–3 pts	
5. Other Metals	0–4 pts	
6. AC Induced Current	0–4 pts	
7. Mechanical Corrosion	0–5 pts	
8. Test Leads	0–6 pts	
9. Close Interval Survey	0–8 pts	
10. Internal Inspection Tool	0–8 pts	
	0–60 pts	60%

Design Index

A.	Pipe Safety Factor	0–20 pts	20%
B.	System Safety Factor	0–20 pts	20%
C.	Fatigue	0–15 pts	15%
D.	Surge Potential	0–10 pts	10%
E.	System Hydrostatic Test	0–25 pts	25%
F.	Soil Movements	0–10 pts	10%
		0–100 pts	100%

Incorrect Operations Index

A. Design
1. Hazard Identification .... 0–4 pts
2. MAOP Potential .... 0–12 pts
3. Safety Systems .... 0–10 pts
4. Material Selection .... 0–2 pts
5. Checks .... 0–2 pts

0–30 pts   30%

B. Construction
1. Inspection .... 0–10 pts
2. Materials .... 0–2 pts
3. Joining .... 0–2 pts
4. Backfill .... 0–2 pts
5. Handling .... 0–2 pts
6. Coating .... 0–2 pts

0–20 pts   20%

C. Operation
1. Procedures .... 0–7 pts
2. SCADA/Communications .... 0–5 pts
3. Drug-testing .... 0–2 pts
4. Safety Programs .... 0–2 pts
5. Surveys .... 0–2 pts
6. Training .... 0–10 pts
7. Mechanical Errors Preventers .... 0–7 pts

0–35 pts   35%

D. Maintenance
  1. Documentation .... 0–2 pts
  2. Schedule .... 0–3 pts
  3. Procedures ....0–10 pts
                                    0–15 pts   15%
        Incorrect Operations Index 0–100 pts   100%

Total Index Sum........................... **0–400 pts**

Leak Impact Factor

A. Product Hazard (Acute + Chronic Hazards) .. 0–22 points
  1. Acute Hazards
      a. N_f... 0–4
      b. N_r... 0–4
      c. N_h 0–4
                    Total ($N_h + N_r + N_f$) 0–12
  2. Chronic Hazard, RQ 0–10
B. Dispersion Factor (Spill Score) ÷ (Population Score) ...0–6
  1. Liquid Spill or Vapor Spill 0–6
  2. Population Density 0–4
Leak Impact Factor = (Product Hazard)/(Dispersion Factor)

Relative Risk Score= (Index sum)/(Leak Impact Factor)
                = 0–400 points

PIPELINE
RISK
MANAGEMENT
MANUAL

CHAPTER ONE

Risk and Quality: Theory and Application

The Role of Entropy

One of Murphy's[1] famous laws states that "left to themselves, things will always go from bad to worse." This humorous prediction is, in a way, echoed in the second law of thermodynamics. That law deals with the concept of *entropy*. Stated simply, entropy is a measure of the disorder of a system. The thermodynamics law states that "entropy must always increase in the universe and in any hypothetical isolated system within it [14]." Practical application of this law says that to offset the effects of entropy, energy must be injected into any system. Without adding energy, the system becomes increasingly disordered.

Although the law was intended to be a statement of a scientific property, it was seized upon by "philosophers" who defined *system* to mean a car, a house, economics, a civilization, or anything that became disordered. By this logic, the concept is universal. It explains why a desk or a garage becomes increasingly cluttered until a cleanup

---

[1] *Murphy's Laws are a famous parody on scientific laws, humorously pointing out all the things that can and often do, go wrong in science and life.*

1

(injection of energy) is initiated. Gases diffuse and mix in irreversible processes; un-maintained buildings eventually crumble; engines (highly ordered systems) breakdown without the constant infusion of maintenance energy.

Another way of saying it is: "Mother Nature hates things she didn't create." Forces of nature seek to disorder man's creations until the creation is reduced to the most basic components. Rust is an example—metal is seeking to disorder itself by reverting to its mineral form.

If we indulge ourselves with this line of reasoning, we may soon conclude that pipeline failures will always occur unless an appropriate kind of energy is applied. Transport of products in a closed conduit, often under high pressure, is a highly ordered, highly structured undertaking. If nature indeed seeks increasing disorder, forces are continuously acting to disrupt this structured process. According to this way of thinking, a failed pipeline with all its product released into the atmosphere or into the ground, or equipment and components decaying and reverting to their original pre-manufactured states represent the less ordered, more natural state of things.

Whether or not this theory can be scientifically proven, it is a useful way of looking at portions of our world. If we adopt a somewhat paranoid view of forces continuously acting to disrupt our systems, we become more vigilant. We take actions to offset those forces. We inject energy into a system to counteract the effects of entropy. In pipelines, this energy takes the forms of maintenance, inspection, patrolling—protecting the pipeline from the forces seeking to tear it apart!

Years of experience in the pipeline industry have established activities that are thought to directly offset specific threats to the pipeline. Such activities include patrolling, valve maintenance, cathodic protection and all the other items discussed in Chapters 3 through 6. Many of these activities have been mandated by governmental agencies. Where the activity was not deemed to be effective in addressing a threat, it was eventually changed or eliminated. This activity list is being refined on a continuing basis.

A logical risk assessment method should follow these same lines of reasoning. All activities that influence, favorably or unfavorably, the pipeline should be considered—even if comprehensive statistical data on the effectiveness of a particular activity is not available. Industry experience and operator intuition can and should be included in the risk picture. Unfortunately, this approach must, from a practical standpoint, often contain an element of subjectivity. So long as this subjectivity is standardized, however, accuracy in a relative risk assessment is not lost.

But, what is risk? Is risk synonymous with hazard?

We define a hazard as a characteristic or group of characteristics that provide the potential for a loss. Flammability or toxicity are examples of such characteristics.

For our purposes, risk is defined as the probability of an event that causes a loss and the magnitude of that loss. Transportation of hazardous products by pipeline is a risk because of the potential of the hazardous product to cause a loss, if it were to be released. The event of releasing the pipeline contents is referred to as a pipeline failure. By this definition, risk is increased when either the probability of the event increases or when the magnitude of the loss (the consequences of the event) increases. The loss is usually defined in economic terms.

It is important to make the distinction between a hazard and a risk. *We can change the risk without changing the hazard.*

When a person crosses a busy street, the hazard should be clear to that person. Loosely defined, it is the prospect that the person must place himself in the path of moving vehicles that can cause him great bodily harm were he to be struck by one or more of them. The hazard is therefore being struck by a moving vehicle. The risk, however, is dependent upon how that person conducts himself in the crossing of the street. He most likely realizes that the risk is reduced if he crosses in a designated, traffic controlled area and takes extra precautions against vehicle operators who may not see him. He has not changed the hazard—he can still be struck by a vehicle—but his risk of injury or death is reduced by prudent actions. Were he to encase himself in an armored vehicle for the trip across the street, his risk would be reduced even further—he has reduced the consequences of the hazard.

The implication of our definition of risk is that the risk is not a static quantity. It can be constantly changing. Along the length of a pipeline, conditions are usually changing. As they change, the risk is also changing in terms of what can go wrong, the likelihood of the event, and the consequences of the event. Because the conditions also change with time, time becomes an indirect factor in the risk. When we perform a risk evaluation, we are actually taking a snapshot of the risk picture at a moment in time.

As was hinted above, a complete risk evaluation requires that three questions be answered:

- What can go wrong?
- How likely is it?
- What are the consequences?

By answering these questions, the risk is defined.

What Can Go Wrong?

Before we can assess the possible contributors to a pipeline failure, we must first define the failure modes. Simply, failure occurs when any portion of the pipeline system allows significant quantities of product to be released unintentionally. "Significant quantities" is included in the definition to distinguish "failure" from nuisance leaks. Unless the product being transported is extremely toxic, the microscopic leaks around flanges or equipment are considered to be inconsequential (for our purposes here).

Most pipeline systems must contain some pressure. This requires a certain strength in the containing structure. If the structure does not have enough strength, failure occurs. Loss of strength can occur because of loss of material thickness from corrosion or from mechanical damage such as scratches and gouges. Failure also occurs if the structure is subjected to stresses beyond its design capabilities. Overpressure, excessive bending, and extreme temperatures are examples.

The answers to the *What can go wrong?* question must be comprehensive. EVERY possible failure mode and initiating cause must be identified. At this stage, the probabilities associated with the failure events are not considered. Even the most remotely possible failure types must be included here. Complex scenarios involving many interlinked events must also be generated. Unexpected interactions between otherwise safe events are often overlooked when hazards are identified.

A powerful tool to use in identifying the hazards is a Hazard and Operability Study (Haz Ops). In this technique, a team of experts is guided through a series of meetings where imaginative scenarios are developed and analyzed by the team. The strength of this technique is the thoroughness of the evaluation. See Battelle Columbus [2] for details.

In this book, the *What can go wrong?* question is addressed by all the items considered in each index. The indexes correspond to historical causes of pipeline failures and the items within each index are those conditions or actions that impact the failure potential. Hazard identification studies (similar to Haz Ops) were used to generate the item list under each index.

How Likely Is It?

Once the hazards have been identified, probabilities of events leading to an accident are calculated. Where several events must happen to initiate the accident, the probabilities of the individual events are combined to arrive at the accident probability. This combination of probabilities may be in series or in parallel, depending upon how the events interact.

Ideally, historical event probabilities would be used here. Historical data, however, is not generally available for all possible event sequences. Furthermore, when data is available, it is normally rare-event data—one failure in many years of service, for instance. Extrapolating future failure probabilities from these small data bases can lead to significant errors. They also imply a false precision, which is discussed under *Subjective Risk Assessments*.

Another possible problem with using historical data is the assumption that the conditions remain constant. For example, when historical data shows a high occurrence of corrosion-related leaks, the operator hopefully takes appropriate action to reduce those leaks. History will foretell the future only when no offsetting actions are taken. While an important piece of evidence, historical data alone should not determine failure probabilities.

The historical rate of failures may tell an evaluator something about the system he is evaluating. Figure 1-1 is a graph that illustrates the well-known "bath tub" shape of many failure rates. For many pieces of equipment or installations, there is a high initial rate of failure. This first portion of the curve is called the "burn-in phase" or the "infant mortality phase." Here, defects developed during initial construction are causing failures. As these defects are eliminated, the curve levels off into the second zone. This is the constant failure zone and reflects the phase where random accidents are maintaining a fairly constant failure rate. Far into the life of the component, the failure rate may

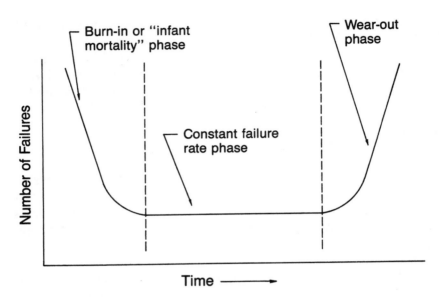

Figure 1-1. Common failure-rate curve (bathtub curve).

begin to increase. This is the zone where things begin to wear out as they reach the end of their useful service life. An overall view of the failure data may suggest such a curve and tell the evaluator what stage the system is in and what can be expected.

In this risk evaluation system, the likelihood of an event occurring is addressed by the relative scoring of the items within each index. The score of each item reflects the importance of that item relative to the other items in the index. This importance is based upon general operator experience including historical failure data, near-misses, and the general knowledge base of the pipeline personnel. Wherever possible, this knowledge base should include experiences from all pipeline operators—not exclusively one company's experiences.

What Are the Consequences?

Inherent in any risk evaluation is a judgment of the consequence factor. We place a value on the consequences of an accident. This then allows us to determine how much we are willing to spend to prevent that accident. Society has only a finite amount of resources to spend on safety and risk reduction. Value judgments as to where resources are to be allocated must involve objective evaluation of the losses that are being avoided by allocating those resources.

Most of the loss components are easy to quantify. In the case of a major pipeline accident (product escaping, perhaps causing an explosion and fire), we could quantify losses such as damaged buildings, vehicles, and other property; costs of service interruption; cost of the product lost; and cost of the cleanup, etc. If lives are lost, however, what value do we place on them? Much has been written on this subject.

There are currently two primary methods used for determining the economic value of a human life. It must be pointed out that this is a "statistical life," not an identified individual. Society has always been willing to spend much more to save an individual in a specific situation—a trapped coal miner, for instance. The statistical life reflects the amount that society is willing to spend to reduce the statistical risk of accidental death by one death.

The first method is the human capital approach in which the value is based upon the economic loss of future contributions to society by an individual. The other approach, willingness to pay, looks at how much an individual is willing to pay (in terms of other goods and services given up) to gain a reduction in the probability of accidental death. Each method has its drawbacks and benefits. While different studies and different federal agencies arrive at different numbers, the EPA implies that $1.5 million is the value of life currently being used. This is the threshold for which substance regulation is determined—regulation is warranted if the cost per life saved does not exceed $1.5 million. Again, this is a statistical life, not a value placed on any identifiable individual [25].

In general, society decides what is an acceptable level of risk for any particular endeavor. What is acceptable for highway traffic deaths is generally not acceptable for pipeline accident deaths, for instance. Many social and economic considerations are thought to influence the human risk tolerance. These are beyond the scope of this text. A main principle, however, is that risk reduction is a cost to society. Society weighs the costs of improved safety in a specific situation against alternate expenditures. Do we spend an extra dollar to spare one traffic fatality every ten years? or do we spend that dollar to feed a hungry child for two days? These types of value judgments help determine the acceptable risk.

An ironic phenomenon may occur in the quest for risk reduction in pipelining. Because most activities are cost-driven, money spent in the name of safety may actually increase the overall risks. For example, if safety-enhancing spending is mandated for pipelines, the increased costs may drive more freight to alternate transportation modes. If these alternate modes are less safe than pipelines, society's risk exposure has actually increased.

Risk Assessment Methods

Battelle Columbus [2] identifies eleven hazard evaluation procedures in common use by the chemical industry. Each has strengths and weaknesses, including costs of the evaluation and appropriateness of an evaluation to a situation. They are listed as:

- checklists
- safety review
- relative ranking
- preliminary hazard analysis
- "what if" analysis
- hazard and operability studies
- failure modes, effects, and criticality analysis
- fault tree analysis
- event tree analysis
- cause-consequence analysis
- human-error analysis

Perhaps the most mathematically rigorous risk assessment method currently in use is the Probabilistic Risk Assessment (PRA). In this approach, every possible failure mechanism and chain of events leading to a failure is studied in great detail. Probabilities are assigned to each component of each event and combined to give a probability for the whole event.

Subjective Risk Assessments

Subjective risk assessments are a special category of risk evaluation techniques, although they overlap many of the above methods. They are done everyday by most of us. A subjective risk assessment occurs when risk is judged not solely upon numerical data. When knowledge is incomplete and opinion, experience, intuition, and other non-quantifiable resources are used, the assessment becomes at least partially subjective.

As operators of motor vehicles, we generally know the hazards associated with driving as well as the consequences of vehicle accidents. At one time or another, most drivers have been exposed to driving accident statistics as well as pictures or graphic commentary of the consequences of accidents. Were we to perform a scientific quantitative risk analysis, we might begin by investigating the accident statistics of the particular make and model of the vehicle we operate. We would also want to know something about the crash survivability of the vehicle. Vehicle condition would also have to be included in our analysis. We might then analyze various roadways for accident history including the accident severity. We would naturally have to compensate for newer roads that have had less opportunity to accumulate an accident frequency base. To be complete, we would have to analyze driver condition as it contributes to accident frequency or severity, as well as weather and road conditions. Some of these variables would be quite difficult to scientifically quantify.

After a great deal of research and using a number of critical assumptions, we may be able to build a system to give us an accident probability number for each combination of variables. For instance, we may conclude that, for vehicle type A, driven by driver B, in condition C, on roadway D, during weather and road conditions E, the accident frequency for an accident of severity F is once for every 200,000 miles driven.

Does this now mean that until 200,000 miles are driven, no accidents should be expected? Does 600,000 miles driven guarantee three accidents? Of course not. All that we know for sure from our study of statistics and probabilities, is that, given a large enough data set, the accident frequency for this set of variables will tend to move towards once every 200,000 miles on average. This may mean an accident every 10,000 miles for the first 100,000 miles followed by no accidents for the next 1,900,000 miles—the average is still once every 200,000 miles.

What we are most interested in, however, is the relative amount of risk to which we are exposing ourselves during a single drive. Our study has told us little about the risk of this drive until we compare this drive to other drives. Suppose we change weather and road conditions to state G from state F and find that the accident frequency is now once every 190,000 miles. This finding now tells us that condition G has *increased* the risk by a small amount. Suppose we change roadway D to roadway H and find that our accident frequency is now once every 300,000 miles driven. This tells us that by using road H we have *reduced* the risk quite substantially compared to using road D (about 50%). Chances are however, we could have made these general statements without the complicated exercise of calculating statistics for each variable and combining them for an overall accident frequency.

So why use numbers at all? Suppose we now make both variable changes simultaneously. The risk reduction obtained by road H is somewhat offset by the increased risk associated with road and weather condition F, but what is the result when we combine a "small risk increase" with a "substantial risk reduction"? As all the variables are subject to change, we need some method to see the overall picture. This requires numbers, but the numbers can be RELATIVE numbers which merely show that variable H has a greater effect on the risk picture than does variable G. Absolute numbers such as the accident frequency numbers used earlier, not only are difficult to obtain, they also give a false sense of precision to the analysis. If we can only be sure of the fact that change X reduces the risk and it reduces it more than change Y does, it is of little further value to say that a once in 200,000 frequency has been reduced to a once in 210,000 frequency by change X and only a once in 205,000 frequency by change Y. We are ultimately most interested in the relative risk picture of change X vs change Y.

The above reasoning forms the basis of the subjective risk assessment. The experts come to a consensus as to how a change in a variable impacts the risk picture, relative to other variables in the risk picture. If real data is available, it is certainly used, but it is used *outside* the risk analysis system. The data is used to help the experts reach a consensus on the importance of the variable and its effects on the risk picture. The consensus is then used in the risk analysis.

The Experts

The term *experts* as it is used here, refers to people knowledgeable in the subject matter. An expert is not restricted to a scientist or other technical person. The experience and intuition of the entire work force should be tapped as much as is practical.

Experts bring to the assessment a body of knowledge that goes beyond statistical data. Experts will discount some data that does not adequately represent the scenario being judged. Similarly, they will extrapolate from dissimilar situations which may have better data available. Every driver who drives in the combination of variables discussed before, has some expertise. Compensating for poor visibility by slowing down demonstrates a simple application of subjective risk assessment. The driver knows that a change in the weather variable of visibility impacts the risk picture in that the driver's reaction times are reduced. Reducing vehicle speed compensates for the reaction time. While this example appears obvious, reaching this conclusion on the basis of statistical data alone would be difficult.

The experience factor and the intuition of experts should not be discounted merely because they cannot be easily quantified. Normally there will be little disagreement among the knowledgeable when risk contributors and risk reducers are evaluated. If there are differences that cannot be resolved, the risk evaluator can merely have each opinion quantified and then produce an average to use in the assessment.

The Scoring System

The risk assessment technique used in this book is best described as a "scoring system." It is a hybrid of several of the methods listed previously and falls into the category of subjective risk assessments. Numerical values (scores) are assigned to conditions on the pipeline system that contribute to the risk picture. The scores are determined from a combination of statistical failure data and operator (expert) experience. The great advantage of this technique is that a much broader spectrum of information can be included—for example, near misses as well as actual failures are considered. The major drawback is the subjectivity of the scoring. Extra efforts must be employed to ensure consistency in the scoring.

As previously mentioned, the score reflects the importance of the item relative to other items. Higher scores mean more importance. More common as well as more catastrophic event items are more important in this sense. So, event probability and consequence are both reflected in the item score. Similarly, more effective preventions are more important—they score higher—than the less effective activities.

The technique of this subjective scoring system is relatively simple and straightforward. The pipeline risk picture is examined in two general parts. The first part is a detailed itemization and relative weighting of all reasonably foreseeable events that may lead to the failure of a pipeline—*What can go wrong?* and *How likely is it?* The second part is an analysis of potential consequences if a failure should occur. This second part addresses the relatively constant *hazard*—its acute and chronic nature. The first part will highlight operational and design options that can change the risk exposure.

The itemization area is further broken into four indexes (Figure 1-2). The indexes roughly correspond to typical categories of reported pipeline accident failures. That is, each index reflects a general area to which, historically, pipeline accidents have been attributed. By considering each item in each index, the evaluator arrives at a numerical value for that index. The four index values are then summed for a total value. This value will be used in the next part when the potential hazards are considered. The individual item values, not just the total index score, are preserved, however, for detailed analysis later.

In the second part, a detailed analysis is made of the potential consequences of a pipeline failure. Product characteristics, pipeline operating conditions, and the line location are considered in arriving at a *consequence factor*. It is called the *Leak Impact Factor* and includes acute as well as chronic hazards associated with product releases. The *Leak Impact Factor* is combined with the index sum (by multiplying) to arrive at a final risk score (Figure 1-2).

This technique is repeated for each section of pipeline. The end result is a numerical risk value for each pipeline section. All the information incorporated into this number is preserved for a detailed analysis, if required.

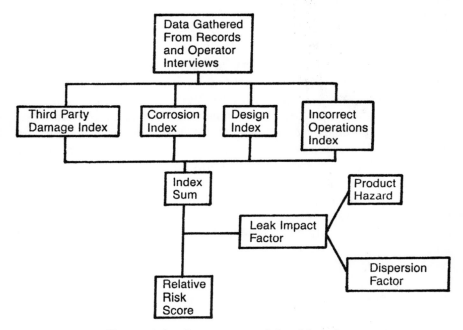

Figure 1-2. Components of the risk rating.

Quality and Risk Management

Quality is a popular term in today's management and industry circles. It implies a way of thinking and a way of doing business. It is widely believed that attention to quality concepts is a requirement to remain in business in today's competitive world markets.

Risk management is basically a method to improve quality. In its best application, it goes beyond basic safety issues to address cost control, planning, and customer satisfaction aspects of quality. For those who link quality with competitiveness and survival in the business world, there is an immediate connection to risk management. In one sense of the word, the prospect of a company failure due to poor cost control or poor decisions, is a risk that can also be managed.

Quality is difficult to precisely define. While several different definitions are possible, they typically refer to concepts such as:

- fitness-for-use
- consistency with specifications
- freedom from defects

with regards to the product or service that the company is producing. All definitions incorporate (directly or by inference) some reference to customers. Broadly defined, a customer is "anyone to whom you provide product, service, or information." Under this definition, almost any exchange or relationship involves a customer. The customer drives the relationship because he specifies what product, service or information he wants and what he is willing to pay for it.

In the pipeline business, typical customers include those who rely upon product movements for raw materials, such as refineries; those who are end users of products delivered, such as residential gas users; and those who are affected by pipelining activities, such as adjacent landowners. As a whole, customers ask for adequate quantities of products to be delivered:

- with no service interruptions (reliability)
- with no safety incidents
- at lowest cost

This is quite a broad brush approach. To be more accurate, the qualifiers of *no* and *lowest* must be defined. Obviously, there are trade-offs involved. Improved safety and reliability may increase costs, and vice versa. Different customers will place differing values on these requirements as was previously discussed in terms of acceptable risk levels.

For our purposes, we can view regulatory agencies as representing the public. The public includes several customer groups with some-times conflicting needs. Those vitally concerned with public *safety* vs those vitally concerned with *costs,* for instance, are occasionally at odds with one another. When a regulatory agency mandates a pipeline safety or maintenance program, this can be viewed as a customer requirement originating from that sector of the public that is most concerned with the safety of pipelines.

As a fundamental part of the quality process, we must make a distinction between types of work performed in the name of the customer.

- **Value-added work.** The work activities that DIRECTLY add value, as defined by the customer, to the product or service. By moving a product from point A to point B, value has been added to that product because it is more valuable (to the customer) at point B than it was at point A.
- **Necessary work.** Work activities that are not value-added, but are necessary in order to complete the value-added work. Protecting the pipeline from corrosion does not directly move the product, but it is necessary in order to ensure that the product movements continue uninterrupted.
- **Waste.** This is the popular name for a category that includes all activities performed which are unnecessary. Repeating a task because it was done improperly the first time is called rework and is included in this category. Tasks that are done routinely, but really do not directly or indirectly support the customer needs are considered to be waste.

The most profitable companies will be those that are most successful in reducing the waste category and optimizing the value-added and necessary work categories. A risk management program is an integral part of this, as will be seen.

Central to many of the quality concepts is the notion of reducing variation. This is the discipline that will ultimately produce the most successful companies.

The simplified process goes something like this: The proper work (value-added and necessary) is identified by studying customer needs and creating the ideal processes to satisfy those needs in the most efficient manner. Once the proper work is identified, the processes that make up that work should be clearly defined and measured. Deviations from the ideal processes are waste. When the company can produce exactly what the customer wants without any variation in that production, that company has gained control over waste in its processes. From there, the processes can be even further improved to reduce costs and increase output, all the while measuring to ensure that variation doesn't return.

This is exactly what risk management should do—identify needs, analyze cost vs benefit of various choices, establish an operating discipline, measure all the processes, and continuously improve all aspects of the operation. Because the pipeline capacity is set by system hydraulics, line size, regulated operating limits, and other fixed constraints, gains in pipelining activities are mostly made by reducing the incremental costs associated with moving the products. These costs are reduced by spending in areas that reap the largest benefits. The chief benefit is the reliability of the pipeline. Spending to prevent losses and service interruptions is an integral part of optimizing pipeline costs.

The pipeline risk items considered in this book are all either conditions or work processes. The conditions are characteristics of the pipeline environment and are not normally changeable. The work processes, however, are changeable and should be directly linked to the conditions. This distinction between conditions and processes is used throughout the text and is described in detail in *Attributes and Preventions* in the next chapter.

The point is that every work process, every activity, even every individual motion should have a purpose. That purpose should be meeting customer requirements. A risk management program should assess each activity in terms of its cost vs benefit. Because every activity and process costs something, it must generate some benefit—otherwise it is waste. Measuring the benefit, including the benefit of loss prevention, allows spending to be prioritized.

Rather than having a broad pipeline operating program to allow for all contingencies, risk management allows the direction of more energy to the areas that need it more. Pipelining activities can be fine-tuned to the specific needs of the various pipeline sections.

A main customer requirement in the pipeline industry is safety. Safety performance in turn directly impacts the other normal customer requirements of reliability and costs. The prudent pipeline operator can gain control over pipeline safety (and hence many other aspects of the operation) by developing and implementing a risk management program. Time and money should be spent in the areas where the return (the benefit) is the greatest. Measurement systems are required to track progress, for without measurements, progress is only an opinion.

The risk evaluation program described here provides a first step in improving the overall quality of a pipeline operation. This is not necessarily because any new techniques are suggested, but rather a discipline is introduced to evaluate all pipeline activities and to score them in terms of their benefit to customer needs. When an extra dollar is to be spent, the risk evaluation program points to where that dollar will do the most good. Dollars presently being spent on one activity may produce more value to the customer if they were being spent another way. Again, the risk evaluation program points this out.

Risk Measurements

At the core of all quality concepts is the need for measurement. *If you don't have a number, you don't have a fact—you have an opinion.* It is always possible to quantify things we truly understand. When we find it difficult to express something in numbers, it is usually because we don't have a complete understanding of the concept. Examples are emotions; we don't try to enumerate love, hate, or anger. We also don't fully understand them.

There exists a discipline to measuring. Before the data gathering effort is started, four questions should be addressed:

1) What will the data represent?
2) How will the values be obtained?
3) What sources of variation exist?
4) Why is the data being collected?

What Will the Data Represent?

In this risk assessment model, the data will represent the relative risk of a section of a pipeline. Inherent in the number will be a complete evaluation of the section's environment and operation. The number will be meaningful only in the context of similar evaluations of other pipeline sections.

How Will the Values Be Obtained?

The data will normally be obtained by an evaluator or team of evaluators who will visit the pipeline operations offices to personally gather the information required to make the assessment. Steps should be taken to ensure consistency in the evaluations. Re-evaluations will be scheduled periodically or it will be left to the operators to update the records.

What Sources of Variation Exist?

Sources of variation include:

- differences in the pipeline section environments
- differences in the pipeline section operation
- differences in the amount of information available on the pipeline section
- evaluator-to-evaluator variation in information gathering and interpretation
- day-to-day variation in the way a single evaluator assigns scores

Why Is the Data Being Collected?

The underlying reason may vary depending upon the user, but hopefully, the common link will be the desire to better understand the pipeline and its risks in order to make improvements in the risk picture. Secondary reasons may include:

- assure regulatory compliance
- set insurance rates
- define acceptable risk levels
- prioritize maintenance spending
- assign dollar values to pipeline systems
- track pipelining activities

Every measurement has a level of uncertainty associated with it. To be correct, a measurement should express this uncertainty: 10 ft $\pm$ 1 in, 15.7°F $\pm$ 0.2°. This uncertainty value represents some of the sources of variations previously listed—operator effects, instrument effects, day-to-day effects, etc. These effects are sometimes called measurement "noise." The variations that we are trying to measure, the relative pipeline risks, are hopefully much greater than the noise. If the noise level is too high relative to the variation of interest, or if the measurement is too insensitive to the variation of interest, the data becomes less meaningful. Wheeler [31] provides detailed statistical methods for determining the "usefulness" of the measurements.

If more than one evaluator is to be used, it is wise to quantify the variation that may exist between the evaluators. This is easily done by comparing scoring by different evaluators of the same pipeline section. The repeatability of the evaluator can be judged by having him/her perform multiple scorings of the same section (this should be done without the evaluator's knowledge that he is repeating a previously performed evaluation).

If these sources of variation are high (>10% of the total score of a section, perhaps), steps should be taken to reduce the variation. These steps may include:

- evaluator training
- refinement of the tool to remove more subjectivity
- changes in the information-gathering activity
- use of only one evaluator

When the scores are obtained, it may be useful to build a picture of the data. Scores may represent a frequency distribution similar to that shown in Figure 1-3. On this distribution, specifications can be

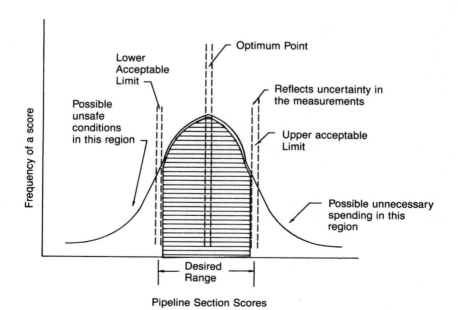

Figure 1-3. Interpreting the data.

noted. They are shown as acceptable limits of the scores. The specifications may change from one pipeline product to another, or between geographical areas. Note that the measurement uncertainty is shown in the specifications. An absolute cutoff point may be unrealistic, if the measurement noise is such that any point selected has a high degree of uncertainty attached to it.

CHAPTER TWO

Risk Assessment Process

Using This Manual

To Get Answers Quick!

This book provides a detailed method of setting up a pipeline relative risk evaluation system. While this topic does fill the pages of this book, the risk assessment tool is not necessarily complex. It can be set up and functioning in a relatively short time by just one evaluator.

A reader need only adopt the standard (suggested) format and point values to begin assessing the risk immediately. The standard format with suggested weightings of all items is shown in *Risk Evaluation at a Glance* (see p. xiv). A risk evaluator with little or no pipeline operating experience could most certainly adopt this approach, at least initially. Similarly, an evaluator who wishes to assess pipelines covering a wide range of services, environments, and operators may wish to use this general approach.

For Pipeline Operators

While the approach described above is the easiest way to get started, this tool becomes even more powerful if the user customizes it. The experienced pipeliner should challenge the point schedules—do they match your operating experience? Read the reasoning behind the schedules—do you agree with it? Invite (or require) input from employees at all levels. Most pipeline operations have a wealth of practical expertise that can be used to fine tune this tool to their unique operating environment.

The point here again is to build a useful tool—one that is regularly used to aid in everyday business and operating decision making, one that is accepted and used throughout the organization. Refer to Chapter One for ideas on evaluating the measuring capability of the tool.

Beginning Risk Management

Building the risk management tool takes four steps:

1. Sectioning. Breaking the pipeline system(s) into sections. Section size is dependent upon how often conditions change and upon the cost of data gathering/maintenance vs the benefit of increased accuracy.
2. Customizing. Deciding on a list of risk contributors and risk reducers and the relative importance of each item on the list.
3. Data gathering. Building the data base by completing an evaluation for each pipeline section.
4. Maintenance. Identifying when and how changes in risk items occur; updating the data base to reflect the changes.

Steps 2 and 3 will most likely be the most costly part of the process. These steps can be time-consuming not only in the hands-on aspects, but also in obtaining the necessary consensus from all key players. The initial consensus often makes the difference between a widely accepted and a partially resisted system. Time and resources spent in Steps 2 and 3 should be viewed as initial investments in a successful risk management tool. Steps 1 and 4 are necessary costs in keeping the tool useful, but should not be very costly to perform.

The risk assessment technique described in this book should become a useful tool for pipeline operators and managers. By providing timely answers to sometimes complicated questions, the tool should become a constant reference point for decision making.

The main thrust of this risk assessment is the risk exposure to the public and how that risk can be effectively managed.

Basic Assumptions

There are a few underlying assumptions built into this model. The user, and especially, the customizer of this manual, should be aware of these.

Independence. Hazards are assumed to be additive but independent. That is, each item that influences the risk picture is considered separately from all other items—it independently influences the risk. The overall risk assessment adds all the independent factors together to get a final number. The final number reflects the "area of opportunity" for a failure because the number of independent factors is directly proportional to the risk.

For example, if Event B can only occur if Event A has first occurred, then Event B is given a lower weighting to reflect the fact that there is a lower probability of both events happening. The risk model does not, however, stipulate that Event B cannot happen without Event A.

Worst Case. The worst case condition for a section governs the point value. For instance, if a 5-mile section of pipeline has 3 ft of cover for all but 200 ft of its length (which has only 1 ft of cover), the section is still rated as if the entire 5-mile length has only 1 ft of cover. The evaluator can work around this by his choice of section breaks.

Relative. Point values are meaningful only in a relative sense. A point score for one pipeline section only shows how that section compares to other scored sections. Higher point values represent increased safety—decreased risk. Absolute risk values are not implied.

Subjective. The example point schedules reflect the author's opinions based upon subjective interpretations of pipeline industry experience as well as personal pipelining experience. The relative importance of each item (this is reflected in the "weighting" of the item) is similarly the author's opinions.

Public. Only dangers and hazards to the general public are of interest here. Risks specific to pipeline operators and pipeline company personnel are not included in this system.

Hazards. Special or exotic failure possibilities are covered only minimally. Sabotage and failure as the result of a secondary failure are examples of events that are only addressed indirectly. Economic risks such as cost of service interruption are also addressed only indirectly. Service interruption is very site specific and can be added to this risk analysis as a consequence of failure if the evaluator is able to quantify this consequence.

Weighting. The weightings of the items, their maximum possible point values, reflect the relative importance of the item. Importance is based upon the item's role in risk contribution or reduction. The four indexes are on equal 0–100 point scales. Because accident history (with regards to cause of failures) is not consistent from one company to another, it does not seem logical to rank one index over another on an accident history basis. No other ranking rationale seems logical, either.

Attributes and Preventions

Because the ultimate goal of the risk assessment is to provide a means of risk management, it is useful to make a distinction between two types of risk components. As it was stated earlier, there is a difference between the hazard and the risk. We can usually do little to change the hazard, but we can take actions to affect the risk. Following this reasoning, the evaluator can categorize each index item as an attribute or a prevention. The attributes correspond loosely to the characteristics of the hazard, while the preventions reflect the risk impacting actions. Attributes reflect the pipeline's environment, while preventions are actions taken in response to that environment. Both impact the risk, but a distinction will be useful.

The term *attributes* used in this sense may be defined as characteristics that are difficult or impossible to change. They are characteristics of the pipeline system over which the operator has little or no control. Most sections of the risk assessment have attributes. Examples of such items that are not routinely changed, and are therefore labeled attributes, include:

- soil characteristics
- type of atmosphere
- product characteristics
- the presence and nature of nearby buried utilities

The other category, *preventions,* includes actions that the pipeline designer or operator can reasonably take to affect the risk picture. Examples of preventions include:

- pipeline patrol frequency
- operator training programs
- right-of-way (ROW) maintenance programs

The above examples of each category are pretty clear-cut. The evaluator should expect to encounter some gray areas of distinction between an attribute and a prevention. For instance, consider the proximity of population centers to the pipeline. In one of the indexes, this impacts the potential for third party damage to the pipeline. This is obviously not an unchangeable characteristic because a re-route of the line is usually an option. But in an economic sense, this characteristic may be unchangeable due to unrecoverable expenses that may be incurred to change the pipeline's location. Another example would be the pipeline depth of cover. To change this characteristic would mean a re-burial or the addition of more cover. Neither of these is an uncommon action, but the practicality of such options must be weighed by the evaluator as he classifies a risk component as an attribute or a prevention.

Figure 2-1 illustrates how some of the risk assessment items are thought to appear on a scale with preventions as one extreme, and attributes as the other. Depth of cover, for example, is a changeable item, but usually at great expense. It is usually thought to lean more towards an attribute and is labelled accordingly.

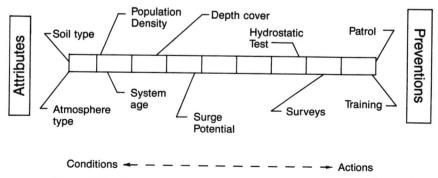

Figure 2-1. Example items on attributes-preventions scale.

Some of these items place the attribute and the prevention in one score. That is, the hazard and specific prevention actions for that hazard are combined into a single score for some items. When this is done, the item is labelled according to which aspect, the attribute or the prevention, plays the larger role. A suggested category will be listed with each item in this text.

As with many aspects of this risk program, consistency is much more important than absolute answers. What is deemed to be an attribute for one section should be an attribute for all sections. Only consistency will allow for meaningful risk comparisons.

The distinction between attributes and preventions is especially useful in risk management policymaking. Company standards can be developed to require certain risk-reducing actions to be taken in response to certain harsh environments. Such a procedure would provide for assigning a level of preventions based upon the level of attributes. The standards can be pre-defined and programmed into a database program to adjust automatically the standards to the environment of the section—harsh conditions require more preventions to meet the standard.

Sectioning of the Pipeline

It must be recognized that, unlike other facilities that undergo a risk assessment, a pipeline usually does not have a constant hazard potential over its entire length. As conditions along the line's route change, so too does the risk picture change. Pipeliners must consider the additional variable: which section of pipeline is being assessed?

The risk evaluator must decide upon a criteria to section the pipeline in order to obtain an accurate risk picture. Breaking the line into many short sections increases the accuracy of the assessment for each section, but may result in higher costs of data collection, handling, and maintenance. Longer sections (fewer in number) on the other hand, may reduce data costs but also reduce accuracy, because average or worst case characteristics must govern if conditions change within the section. A random method of sectioning, such as every mile or between block valves, does not take advantage of obvious break points. Randomizing may actually reduce accuracy and increase costs if inappropriate and unnecessary break points are chosen.

The most appropriate criteria for sectioning is to insert a break point wherever significant changes occur. A significant condition change must be determined by the evaluator with consideration given to data costs and desired accuracy.

The evaluator should scan Chapters 3–7 of this text to get a feel for the types of conditions that make up the risk picture. He should note those conditions that are most variable in the pipeline system being studied and rank those items with regard to magnitude of change and frequency of change. This is probably best done with the input of the pipeline operators. The employee input not only helps to ensure completeness of the assessment, but also helps to build employee acceptance of the risk management technique. This ranking will be subjective and perhaps incomplete, but it will serve as a good starting point for sectioning the line(s). An example of a short list of ranked conditions is as follows:

1. population density
2. soil conditions
3. coating condition
4. age of pipeline

In this example, the evaluator(s) foresee the most significant changes along the pipeline route to be population density, followed by varying soil conditions, then coating condition, and pipeline age. This list was designed for an aging 60-mile pipeline in Louisiana passing close to several rural communities and alternating between marshlands (clay) and sandy soil conditions fits this example list. Furthermore, the coating may be in various states of deterioration (maybe roughly corresponding to the changing soil conditions) and the line has had sections replaced with new pipe over the last few years. These last facts would account for the evaluator choosing coating and age conditions as key changes to be expected along the pipeline route.

Next, the evaluator should insert break points for the sections based upon the top items on the prioritized list of condition changes for a trial sectioning of the line(s). If the number of sections resulting from this process is deemed to be too large, the evaluator needs to merely reduce the list (eliminating conditions from the bottom of the prioritized list) until an appropriate number of sections is obtained. This trial-and-error process is repeated until a cost effective sectioning has been completed.

Example:

Following this philosophy, suppose that the evaluator of this hypothetical Louisiana pipeline decides he will section the line according to the following rules he has developed:

1. Insert a section break each time the population density along a one-mile section changes by more than 10%. These population section breaks will not occur more often than each mile, and as long as the population density remains constant, a section break is unwarranted.

2. Insert a section break each time the soil corrosivity changes by 30%. In this example, data is available showing average soil corrosivity for each 500-ft section of line. Therefore, section breaks may occur a maximum of ten times (5,280 ft per mile divided by 500-ft sections) for each mile of pipeline.

3. Insert a section break each time the coating condition changes significantly. This will be measured by the corrosion engineer's assessment. Because this assessment is subjective and based on sketchy data, such section breaks may occur as often as every mile.

4. Insert a section break each time a difference in age of the pipeline is seen. This is measured by comparing the installation dates. Over the total length of the line, six new sections have been installed to replace unacceptable older sections.

Following these rules, the evaluator finds that his top listed condition causes 15 sections to be created. By applying the second condition rule, he has created an additional 8 sections, bringing the total to 23 sections. The third rule yields an additional 14 sections and the fourth causes an additional 6 sections. This brings the total to 43 sections in the 60-mile pipeline.

The evaluator can now decide if this is an appropriate number of sections. As previously noted, factors such as the desired accuracy of the evaluation and the cost of the data gathering and analysis should be considered. If he decides that 43 sections is too many for the company's needs, he can reduce the number of sections by first eliminating the additional sectioning caused by application of his fourth rule. Elimination of these 6 sections caused by age differences in the pipe is appropriate because it had already been established that this was a lower priority item. That is, it is thought that the age differences in the pipe is not as significant a factor as the other conditions on the list.

If the section count (now down to 37) is still too high, the evaluator can eliminate or reduce sectioning caused by his third rule. Perhaps combining the corrosion engineer's "good" and "fair" coating ratings would reduce the number of sections from 14 to 8.

In the above example, the evaluator has roughed out a plan to break down the pipeline into an appropriate number of sections. Most likely, a target section length was in his mind to begin with. As a starting point, one-mile sections MAY be appropriate in many applications. This is the section length used by the Department of Transportation (DOT) in definitions for population density assessments along the pipeline route. Population density, of course, plays a major role in the risk picture. As with the example, however, one-mile or two-mile sections may be too many for the purposes of the risk assessment. Again, section length involves a trade-off between accuracy and cost.

Figure 2-2 illustrates a piece of pipeline being sectioned based upon population density and soil conditions.

For many items in this evaluation (especially in the *Incorrect Operations Index*) section lines will not have an impact. Items such as training or procedures are generally applied uniformly across the entire

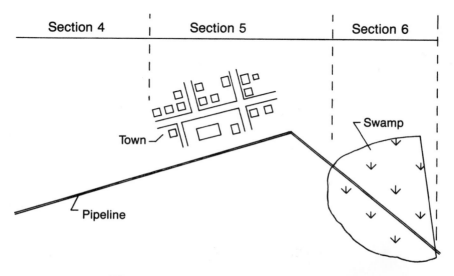

Figure 2-2. Sectioning of the pipeline.

pipeline system or at least within a single operations area. This should not be taken for granted, however, during the data gathering step.

Third Party Damage Index

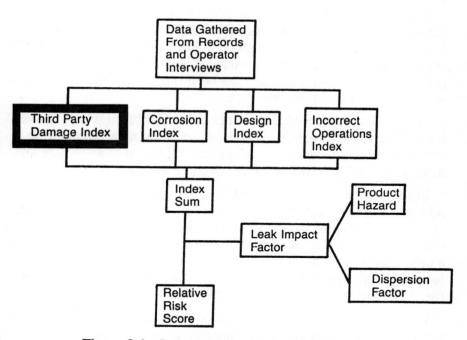

Figure 3-1. Components of risk rating flowchart.

Third Party Risk

A. Minimum Depth of Cover .... 0–20 pts   20%   (p. 34)
B. Activity Level .... 0–20 pts   20%   (p. 38)
C. Aboveground Facilities .... 0–10 pts   10%   (p. 42)

D.	One-Call System.................... 0–15 pts	15%	(p. 44)
E.	Public Education .... 0–15 pts	15%	(p. 46)
F.	Right-of-Way Condition .... 0–5 pts	5%	(p. 48)
G.	Patrol Frequency .... 0–15 pts	15%	(p. 50)
	0–100 pts	100%	

Pipeline operators must take steps to reduce the possibility of damage to their facilities by other people. The extent to which steps are necessary is dependent upon how readily the system can be damaged and how often the chance for damage occurs.

Department of Transportation (DOT) pipeline accident statistics indicate that third party intrusions are the leading cause of pipeline failure. Forty percent of all pipeline failures between 1971 and 1986 are attributed to third party damages. In spite of these statistics, the potential for third party damage has been one of the least considered aspects of pipeline hazard assessment.

The good safety record of pipelines can be attributed in part to their initial installation in sparsely populated areas and their burial 2.5 to 3 feet deep. Today however, development is threatening to intrude and increase the risk of pipeline failure due to excavation damage.

In the period from 1983 through 1987, eight deaths, twenty-five injuries, and over \$14 million in property damage occurred in the hazardous liquid pipeline industry due solely to excavation damage by others. These types of pipeline failures represent 259 accidents out of a total of 969 accidents from all causes. This means that 26.7% of all hazardous liquid pipeline accidents were caused by excavation damage. (See U.S. Dept. of Transportation [29].)

In the gas pipeline industry, a similar story emerges; 430 incidents from excavation damage were reported in the 1984–1987 period. These accidents resulted in 26 deaths, 148 injuries, and over \$18 million in property damage. Excavation damage is thought to be responsible for 10.5% of incidents reported for distribution systems, 22.7% of incidents reported for transmission/gathering pipelines, and 14.6% of all incidents in gas pipelines. (See U.S. Dept. of Transportation [29].)

The pipeline designer and, perhaps to an even greater extent, the operator can affect the risk from third party activities. As an element of the total risk picture, the probability of third party damage to a facility is dependent upon:

- the nature of possible intrusions
- the ease with which the facility can be reached by the intruding party
- the activity level

Possible intruders include:

- excavating equipment
- projectiles
- vehicular traffic
- trains
- farming equipment
- seismic charges
- fence posts
- telephone posts
- anchors
- dredges

Factors that affect the susceptibility of the facility include:

- depth of cover
- nature of cover (earth, rock, concrete, paving, etc.)
- manmade barriers (fences, barricades, levees, ditches, etc.)
- natural barriers (trees, rivers, ditches, rocks, etc.)
- presence of pipeline markers
- condition of right-of-way
- frequency and thoroughness of patrolling
- response time to reported threats

The activity level is judged by items such as:

- population density
- construction activities nearby
- proximity and volume of rail or vehicular traffic
- offshore anchoring areas

- volume of one-call system reports
- number of buried utilities in the area

Serious damage to a pipeline is not limited to actual punctures of the line. A mere scratch on a coated steel pipeline damages the corrosion-resistant coating as a minimum. Such damage can lead to accelerated corrosion and ultimately a corrosion failure perhaps years in the future. If the scratch is deep enough to have removed even a tiny bit of metal, a stress concentration area (see *Design Index*) could be formed which again, perhaps years later, may lead to a failure from fatigue, either alone or in combination with some form of stress corrosion cracking.

This is one reason why public education plays such an important role in damage prevention. To the casual observer, a minor dent or scratch in a steel pipeline may appear insignificant—certainly not worthy of mention. A pipeline operator knows the potential hazard of any disturbance to the line. This hazard should be communicated to the general public.

A. Minimum Depth of Cover   Suggested weighting......... 20%
Attribute

This is the amount of earth cover over the shallowest piece of pipeline—no matter how short that piece may be. Averaging of depths is discouraged. The greatest exposure to potential damage exists where the line has the least amount of cover, regardless of the depth elsewhere. In cases where depth of cover varies, the evaluator may wish to divide the line into sections accordingly.

This item is normally considered to be an attribute because it is not an easily changed condition along the line.

A schedule or simple formula should be developed to assign point values based upon depth of cover:

(amount of cover in inches) ÷ 3 = point value
up to a maximum of 20 points

So: 42 in. of cover = 42 ÷ 3 points = 14 points
    24 in. of cover = 24 ÷ 3 points = 8 points

Points should be assessed based upon the shallowest location within the section being evaluated. The evaluator should feel confident that the depth of cover data is current; otherwise, the point assessments should reflect the uncertainty. (Note that increasing points indicates a safer condition. This is the convention which is used throughout this book.)

Experience tells us that less than one foot of cover may actually do more harm than good. It is enough cover to conceal the line but not enough to protect the line from even shallow earth moving equipment (such as farming equipment). Three feet of cover is the normal amount of cover required by DOT.

The main benefit of earth cover is to protect the line from third party activities that may harm it. Consequently, credit should be given for other means of protecting the line from mechanical damage. A schedule should be developed for these other means. A simple way to do this is to equate the mechanical protection to an amount of additional earth cover.

2 in. concrete coating =   8 in. of additional earth cover
4 in. concrete coating = 12 in. of additional earth cover

Pipe casing = 24 in. of additional cover
Concrete slab (reinforced) = 24 in. of additional cover

For example, by using the example formula above, a pipe section that has 14 in. of cover and is encased in a casing pipe would have an equivalent earth cover of $14 + 24 = 38$ in., yielding a point value of $38 \div 3 = 12.7$.

Burial of a highly visible strip of material with warnings clearly printed may help to avert damage to the pipeline (Figure 3-2). Such flagging or tape is commercially available and is usually installed just beneath the ground surface directly over the pipeline. Hopefully, an excavator will discover the warning tape, cease the excavation, and avoid damage to the line. While this early warning system provides no physical protection, its benefit from a risk standpoint can also be equated to an additional amount of earth cover:

Warning Tape = 6 in. of additional cover

As with all items in this system, the evaluator should use his company's best experience in creating his schedule and point values. Common situations that may need to be addressed include: rocks in one region, sand in another (is the protection value equivalent?);

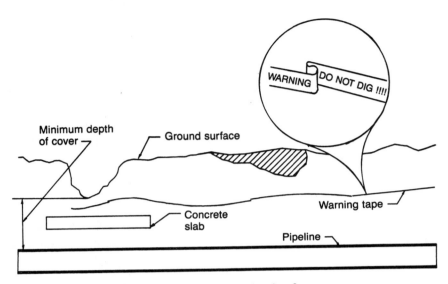

Figure 3-2. Minimum depth of cover.

pipelines under roadways (concrete vs asphalt vs compacted stone, etc.). The evaluator need only remember the goal of consistency and the intent of assessing the amount of existing protection from mechanical damage.

If the wall thickness is greater than what is required for anticipated pressures and external loadings, the extra thickness is available to provide additional protection from external damage or corrosion. Mechanical protection that may be available from extra pipe wall material is accounted for in the *Design Index* section of this book.

In the case of submerged pipelines, the intent is the same. Credit should be given for water depth, concrete coatings, subsea covering of the line, extra damage protection coatings, etc.

A point schedule for submerged lines might look something like this:

Navigable waterways:
Depth below water surface
0–5 ft .... 0 pts
5 ft–max anchor depth .... 3 pts
>max anchor depth .... 7 pts

Depth below bottom of waterway (add these points to the points from Depth below water surface)

0–2 ft .... 0 pts
2–3 ft .... 3 pts
3–5 ft .... 5 pts
5 ft–max dredge depth .... 7 pts
>max dredge depth .... 10 pts

Concrete coating (add these points to the points assigned for water depth and burial depth)

None .... 0 pts
Min. 1 in. .... 5 pts

Note that the point scale for concrete coating of offshore pipe is different from that of onshore pipe.

The total for all three point categories may not exceed 20 pts.

The above schedule assumes that water depth offers some protection against third party damage. This may not be a valid assumption in every case—it must be decided by the evaluator. Such schedules might also reflect the anticipated sources of damage. If only small boats anchor in the area, perhaps a large diameter line is in less danger of immediate damage than a small diameter line. In this case, extra points may be awarded for larger diameters. Reported depths must reflect the current situation as sea or riverbed scour often quickly reduces the cover.

The use of water crossing surveys to determine the condition of the line, especially the extent of its exposure to third party damage, indirectly impacts the risk picture (Figure 3-3). Such a survey may be the only way to establish the pipeline depth and extent of its exposure to boat traffic, currents, floating debris, etc. Because conditions can

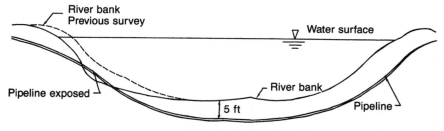

Figure 3-3. River crossing survey.

change dramatically when flowing water is involved, the time since the last survey is also a factor to be considered. Such surveys are also considered in the *Incorrect Operations Index.*

Example applications:

1. In this example, a pipeline section has burial depths of 10 in. and 30 in. In the shallowest portions, a concrete slab has been placed over and along the length of the line. The 4-in. slab is 3 ft wide and reinforced with steel mesh. Using the above schedule, the evaluator calculates points for the shallow sections with additional protection and for the sections buried with 30 in. of cover. For the shallow case: 10 in. cover + 24 in. additional (equivalent) cover due to slab = (10 + 24)/3 pts = 11.3 pts. Second case: 30 in. of cover = 30/3 = 10 pts. Because the minimum cover (including extra protection) yields the higher point value, he uses the 10 pts score of the pipe buried with 30 in. of cover as the worst case and hence, the governing point value for this section.

2. In this section, a submerged line lies unburied on a river bottom, 30 ft below the surface at the river midpoint, rising to the water surface at shore. At the shoreline, the line is buried with 36 in. of cover. The line has 4 in. of concrete coating around it throughout the entire section.

   Points are assessed as follows:

   The shore approaches are very shallow; although boat anchoring is rare, it is possible. No protection is offered by water depth, so 0 pts are given here. The 4 in. of concrete coating yields 5 pts. Because the pipe is not buried beneath the river bottom, 0 pts are awarded for cover.

   Total score = 0 + 5 + 0 = 5 pts.

B. Activity Level   **Suggested weighting......... 20%**
                                 Attribute

Fundamental to any risk assessment is the "area of opportunity." For an analysis of third party damage potential, the area of opportunity is strongly affected by the level of activity near the pipeline. It is intuitively obvious that more digging activity near the line increases the opportunity for a line strike.

DOT accident statistics for gas pipelines indicate that, in the 1984–1987 period, 35% of excavation damage accidents occurred in Class 1 and 2 locations. (See U.S. Dept. of Transportation [29].) These are the less populated areas. This tends to support the assumption that more population means more accident potential.

The activity level item is also normally an attribute because a relocation is usually the only means for the pipeline operator to affect it. A relocation is not a routine option.

The evaluator should create several classifications of activity level. He does this by describing *sufficient* conditions to categorize an area into one of his classifications. The following example provides a sample of some of the conditions that may be appropriate. Further explanation follows the example classifications.

High Activity Level Area. (0 points) This area is characterized by one or more of the following:

- Class 3 population density (as defined by DOT CFR part 192)
- High population density
- Frequent construction activities
- High volume of one-call or reconnaissance reports (>2 per week)
- Rail or roadway traffic posing a threat
- Many other buried utilities nearby
- Normal anchoring area when offshore
- Dredging near the offshore line is common

Medium Activity Level. (8 points) This area is characterized by one or more of the following:

- Class 2 population density (as defined by DOT)
- Low population density nearby
- No routine construction activities that could pose a threat
- Few one-call or reconnaissance reports (<5 per month)
- Few buried utilities nearby

Low Activity Level. (15 points) This area is characterized by *all* of the following:

- Class 1 population density (as defined by DOT)
- Rural, low population density
- Virtually no activity reports (<10 per year)
- No routine harmful activities in the area (Agricultural activities where the equipment cannot penetrate to within 1 ft of the pipeline depth may be considered harmless.)

None. (20 points) The maximum point level is awarded when there is virtually no chance of any digging or other harmful activity near the line.

The evaluator may assign point values between these categories, but should take efforts to ensure consistency.

In each classification of the above example, population density is a factor. More people in an area generally mean more activity; fence building, gardening, water well construction, ditch digging or clearing, wall building, shed construction, . . . the list goes on. Many of these activities could disturb a buried pipeline.

The disturbance could be so minor as to go unreported by the offending party. As already mentioned, such unreported disturbances as coating damage or a scratch in the pipe wall are often the initiating condition for a pipeline failure sometime in the future.

An area that is being renovated or is experiencing a growth phase will require frequent construction activities. These may include soil investigation borings, foundation construction, installation of buried utilities (telephone, water, sewer, electricity, natural gas), and a host of other potentially damaging activities.

Perhaps one of the best indicators of the activity level is the frequency of reports. These reports may come from direct observation by pipeline personnel, patrols by air or ground, and telephone reports by the public or by other construction companies. The one-call systems (these are discussed in a later section), where they are being used, provide an excellent database for assessing the level of activity.

The presence of other buried utilities logically leads to more frequent digging activity as these systems are repaired, maintained, and inspected. This is another measure that can be used in judging the activity level.

Anchoring, fishing, and dredging activities pose the greatest third party threats to submerged pipelines. To a lesser degree, new construction by open-cut or directional-drill methods may also pose a threat to existing facilities. Dock and harbor constructions and perhaps even offshore drilling activities may also be a consideration.

Seismograph Activity. Of special note here is seismograph work. As a part of exploratory work, usually searching for oil or gas reservoirs, energy is transmitted into the ground and measured to determine information about the underlying geology of the area. This usually involves crews laying shot lines—rows of buried explosives that are later detonated. The detonations supply the energy source to gather the information sought. Sometimes, instead of explosive charges, other techniques that impart energy into the soil are used. Examples include a weight dropped onto the ground where the resulting shock waves are monitored and a vibration technique that generates energy waves in certain frequency ranges.

Seismograph activity can be hazardous to pipelines. The first hazard occurs if holes are drilled to place explosives. Such drilling can place the pipeline in jeopardy. Depth of cover provides little protection because the holes are drilled to any depth. The second hazard is the shock waves to which the pipeline is exposed. When the explosive(s) is detonated, a mass of soil is accelerated [11]. If there is not enough backup support for the pipeline, the pipe itself absorbs the energy of the accelerating soil mass [11]. This adds to the pipe stresses (Figure 3-4). Conceivably, a charge (or line of charges) detonated far below the pipeline can be more damaging than a similar charge placed closer

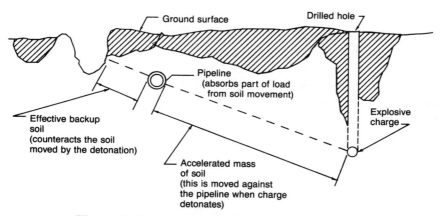

Figure 3-4. Seismograph activity near pipelines.

to the line but at the same depth. An analysis must be performed on a case-by-case basis to determine the extent of the hazard.

As of this writing, pipeline operators have little authority in specifying minimum distances for seismograph activity. Technically, the operator can only forbid activity on the few feet of ROW that he controls. Cooperation from the seismograph company is often sought.

As a component of the risk picture, the potential for seismograph activity near to the pipeline should be evaluated.

C. Aboveground Facilities   Suggested weighting .... 10% Attribute

This is a measure of the susceptibility of aboveground facilities to third party disturbance. A governing assumption here is that aboveground pipeline components have a third party damage exposure as do the buried sections. Contributing to this exposure are the threats of vehicular collision and vandalism. The argument can be made that these threats are partially offset by the benefit of having the facility in plain sight, helping to avoid damages caused by not knowing exactly where the line is (as is the case for buried sections). It is left to the evaluator to adjust the weighting factor and the point schedule to values consistent with the company's judgment and experience.

Although this item is generally considered to be an attribute, this is one of the gray areas of the attribute/prevention distinction. While the presence of aboveground components is something that is often difficult or impossible to change, there are many preventive measures that can be taken to reduce the risk exposure. This item, then, combines the changeable and non-changeable aspects into a single point value.

The evaluator should set up a point schedule that gives the maximum point value for sections with no aboveground components. For sections that do have aboveground facilities, point credits should be given for conditions that lessen the risk of third party damage (Figure 3-5). These conditions will often take the form of vehicle barriers or other barriers or discouragements to intrusion.

No aboveground facilities .... 10 pts
Aboveground facilities .... 0 pts
plus any of the following that apply
(total not to exceed 10 pts)
Facilities more than 200 ft from vehicles .... 5 pts

Area surrounded by 6 ft chainlink fence .... 2 pts
Protective railing (4 in. steel pipe or better) .... 3 pts
Trees (12 in. diameter), wall, or other substantial
    structure(s) between vehicles and facility .... 4 pts
Ditch (minimum 4 ft depth/width) between vehicles and
    facility .... 3 pts
Signs (Warning, No Trespassing, Hazard, etc.) .... 1 pt

Credit may be given for security measures that are thought to reduce vandalism (intentional third party intrusions). The example above allows a small amount of points for signs that may discourage the casual mischief-maker or the passing hunter taking target practice. Lighting, barbed wire, video surveillance, sound monitors, motion sensors, alarm systems, etc., may warrant point credits as risk reducers.

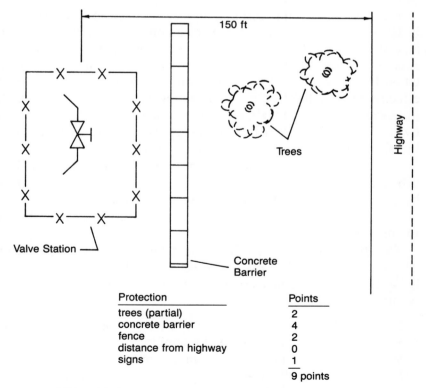

Protection	Points
trees (partial)	2
concrete barrier	4
fence	2
distance from highway	0
signs	1
	9 points

Figure 3-5. Protection for aboveground facilities.

D. One-call Systems

Suggested weighting 15%
Prevention

A one-call system is a service that receives notification of upcoming digging activities and in turn notifies owners of potentially affected underground facilities. A conventional one-call system is defined by the DOT as "a communication system established by two or more utilities (or pipeline companies), governmental agencies, or other operators of underground facilities to provide one telephone number for excavation contractors and the general public to call for notification and recording of their intent to engage in excavation activities. This information is then relayed to appropriate members of the one-call system, giving them an opportunity to communicate with excavators, to identify their facilities by temporary markings, and to follow-up the excavation with inspections of their facilities." Such systems can also be established by independent entrepreneurs.

The first modern one-call system was installed in Rochester, NY, in 1964. As of this writing, there are 88 one-call systems in 47 states and Washington D.C. plus similar systems operating in Canada, Australia, Scotland, and Taiwan. A report by the National Transportation Board on a study of 16 one-call centers gives evidence of the effectiveness of this service in reducing pipeline accidents. In ten instances (of the 16 studied), excavation-related accidents were reduced by 20%–40%. In the remaining six cases, these accidents were reduced by 60%–70%. (see "One-call Systems" [21].)

One-call systems operate within stated boundaries, usually in urban areas. They are the most effective in states that mandate by law that all excavators use the service. Even when the one-call system is voluntary, many major buried utility owners subscribe to the service.

The effectiveness of a one-call system depends upon several factors. The evaluator should assess this effectiveness for the pipeline section being evaluated. A sample point schedule (with explanations following) would be:

Mandated by law .... 4 pts
Proven record of efficiency and reliability .... 2 pts
Widely advertised and well known in community .... 2 pts
Meets minimum ULCCA standards .... 2 pts
Appropriate reaction to calls .... 5 pts

Add points for all applicable characteristics. The best one-call system is characterized by all the above factors and will have a point value of 15 points.

The first category is straightforward; a system mandated by law will be more readily accepted and utilized. The next two point categories are more subjective. The evaluator is asked to judge the effectiveness and acceptance of the system. The one-call service can often provide data showing numbers of reports in each area covered. This data can perhaps be compared to suspected activity levels in the area. Actual reports that equal or exceed the suspected activity level would be a favorable reflection of the system (or an unfavorable reflection of the estimate?). The degree of community acceptance can be assessed by a spot check of local excavators and by the level of advertising of the system. The evaluator may set up a more detailed point schedule to distinguish between differences he perceives. This detailed schedule could be tied to the results of a random sampling of the one-call system.

Another category in this schedule refers to standards established by the Utility Location and Coordination Council of America (ULCCA) for one-call centers. The evaluator may substitute any other appropriate industry standard. This may overlap the first question of whether the one-call system is mandated by law. If mandated, certain minimum standards will doubtlessly be established. Minimum standards may address:

- hours of operation
- record keeping
- method of notification
- off-hours notification systems
- timeliness of notifications

The last category deals with the pipeline company's response to a report of third party excavation activity. Obviously, reports that are not properly addressed in a timely manner negate the effects of reporting. The evaluator should look for evidence that all reports are investigated in a timely manner. A sense of professionalism and urgency should exist among the responders. Appropriate response may include:

- Dispatching of personnel to the site to provide detailed markers of pipeline location
- Pre-job communications or meetings with the excavators
- On-site inspection during the excavation

Inspection of the pipeline facilities after the excavation

The evaluator may look for documentation or other evidence to satisfy himself that an appropriate number of these most critical actions is being taken.

The points are added to get a value for One-Call Systems. A section that is not participating in such a program would get zero points.

E. Public Education Program   Suggested weighting 15%
Prevention

Public education programs play a significant role in reducing third party damage to pipelines. It is thought that most third party damage is unintentional and due to ignorance. This is ignorance not only of the buried pipeline's exact location, but also ignorance of the above-ground indications of the pipeline's presence and ignorance of pipelines in general. A pipeline company committed to educate the community on pipeline matters will almost assuredly reduce its exposure to third party damage.

Some of the characteristics of an effective public education program are shown on the following schedule. A relative point scale is included and more explanation follows the table.

Mailouts .... 2 pts
Meetings with public officials once per year .... 2 pts
Meetings with local contractors/excavators
    once per year .... 2 pts
Regular education programs for community groups ..... 2 pts
Door-to-door contact with adjacent residents .... 4 pts
Mailouts to contractors/excavators .... 2 pts
Advertisements in contractor/utility publications
    once per year .... 1 pt

Add points for all characteristics that apply. The best public education program will score 15 points here.

Regular contact with property owners and residents who live adjacent to the pipeline is thought to be the first line of defense in public education. When properly motivated, these people actually become protectors of the pipeline. They realize that the pipeline is a neighbor whose fate may be closely linked to their own. They may also act as good neighbors out of concern for a company that has taken the time to explain to them the pipeline's service and how it relates to them. Although it is probably the most expensive approach, door-to-door contact is arguably unsurpassed in effectiveness. This is especially true today when pleasant face-to-face contact between large corporations and John Q. Citizen is rare. The door-to-door contact, when performed once per year, rates the highest points in the example schedule.

Other techniques that emphasize the good neighbor approach include regular mailouts, presentations at community groups, and advertisements. Mailouts generally take the form of an informational pamphlet and perhaps a promotional item such as a magnet, calendar, memo pad, pen, rain gauge, tape measure, or key chain with the pipeline company's name and 24-hour phone number. The pamphlet may contain details on pipeline safety statistics, the product being transported, and how the company ensures the pipeline integrity (patrols, cathodic protection, etc.). Most important perhaps, is information that informs the reader of how sensitive the line can be to damage from third party activities. Along with this is the encouragement to the reader to notify the pipeline company if any potentially threatening activities are observed. The other tokens often included in the mailout merely serve to attract the reader's interest and to keep the company's name and number handy.

Mailouts can be effectively sent to landowners, tenants, other utilities, excavation contractors, general contractors, emergency response groups, and local and state agencies.

Professional, entertaining presentations are always welcomed at civic group meetings. When such presentations can also get across a message for public safety through pipeline awareness, they are doubly welcomed. These activities should be included in the point schedule. Any regular advertisements aimed at increasing public awareness of pipeline safety should similarly be included in the schedule.

Meetings with public officials and local contractors serve several purposes for the pipeline operator. While advising these people of pipeline interests (and the impact on THEIR interests), a rapport is established with the pipeline company. This rapport can be valuable in terms of early notification of government planning, impending project work, emergency response, and perhaps a measure of consideration and benefit of the doubt for the pipeline company. Points should be given for this activity to the extent that the evaluator sees the value of the benefits in terms of risk reduction.

Advertising can be company specific or can represent common interests of a number of pipeline companies. Either way, the value is obtained as the audience is made aware or reminded of their role in pipeline safety.

F. Right-of-way Condition   Suggested weighting .... 5% Prevention

This item is a measure of the recognizability and inspectability of the pipeline corridor. A clearly marked, easily recognized right-of-way (ROW) reduces the susceptibility of third party intrusions and aids in leak detection (ease of spotting vapors or dead vegetation from ground or air patrols) (Figure 3-6).

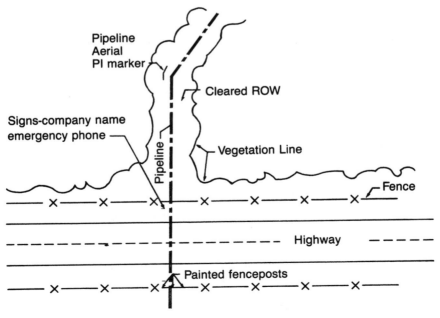

Figure 3-6. Typical pipeline ROW markings.

The evaluator should establish a point schedule with clear parameters. The user of the schedule should be able to tell exactly what actions will increase the point value. The less subjective the schedule, the better, but simplicity is also encouraged. The following example schedule is written in paragraph form where interpolations between paragraph point values are allowed.

Excellent .... 5 pts

Clear and unencumbered; route clearly indicated; signs and markers visible from any point on ROW or from above; signs and markers at all roads, railroads, ditches, water crossings; all changes of direction are marked; air patrol markers are present.

Good .... 3 pts

Clear route (no overgrowth obstructing the view along the ROW from ground level or above); well marked but markers are not visible from every point of ROW or above; signs and markers at all roads, railroads, ditches, water crossings.

Average .... 2 pts

ROW not uniformly cleared; more markers are needed for clear identification at roads, railroads, waterways.

Below Average .... 1 pt

ROW is overgrown by vegetation in some places; ground is not always visible from the air or there is not a clear line of sight along the ROW from ground level; indistinguishable as a pipeline ROW in some places; poorly marked.

Poor .... 0 pts

Indistinguishable as a pipeline ROW; no (or inadequate) markers present.

Select the point values corresponding to the closest description of the actual ROW conditions observed in the section.

Descriptions such as these above should provide the operator with enough guidance to take corrective action. Point values can be more specific (markers at 90% of road crossings......2 pts; at 75% of road crossings.......1 pt; etc.) but this may be an unnecessary complication.

G. Patrol Frequency

Suggested weighting 15%
Prevention

Patrolling the pipeline is a proven effective method of reducing third party intrusions. The frequency and effectiveness of the patrol should be considered in assessing the patrol value.

Patrolling becomes more necessary as more third party activities are unreported. The amount of unreported activity will depend upon many factors, but one source (see Bolt and Logtenberg [3]) reports the number of unreported excavation works around a pipeline system in the U.K. to be 25% of the total number of excavation works. This is estimated to be around 775 unreported per year on their 10,400 km system [3]. While unreported excavation does not automatically translate into pipeline damage, obviously the potential exists that some of those 775 excavations will contact the pipeline.

From a reactive standpoint, the patrol is also intended to detect evidence of a leak such as vapor clouds, unusual dead vegetation, bubbles from submerged pipelines, etc. As such, it is a proven leak detection method (see *Leak Impact Factor*).

From a proactive standpoint, the patrol also should detect impending threats to the pipeline. Such threats take the form of excavating equipment operating nearby, new construction of buildings or roads, or any other activities that could cause a pipeline to be struck, exposed, or otherwise damaged. Note that some activities are only indirect indications of threats. A building several hundred yards from the pipeline will not pose a threat in itself, but the experienced observer will investigate where supporting utilities are to be directed. Construction of these utilities at a later time may create the real threat.

The patrol should also seek evidence of activity that has already passed over the line. Such evidence is usually present for several days after the activity and may warrant inspection of the pipeline.

A direct measure of the patrol effectiveness would be data showing a number of situations that were missed by the observers when the opportunity was there. Indirect measures include observer training and analysis of the detection opportunity. This opportunity analysis would look at altitude and speed of aerial patrol and, for ground patrol, perhaps the line of sight along and either side of the ROW. In both methods, the opportunity for early discovery lies in the ability to detect activities before the pipeline ROW is encroached.

The suggested point schedule will award points based on patrol frequency under the assumption of optimum effectiveness. If the evaluator judges the effectiveness to be less, he can reduce the points to the equivalent of a lower patrol frequency. This is reasonable because lower frequency and lower effectiveness both reduce the area of opportunity for detection.

Whenever possible, the patrol frequency should be determined based upon a statistical analysis of data. Historical data will often follow a typical rare-event frequency distribution such as Figure 3-7. Once the distribution is approximated, analysis of the curve will enable some predictive decisions to be made. Because the data will often be rare occurrence events, a management decision of "acceptable risk" will be needed. For example, management may decide that the appropriate patrol frequency should detect, with a 90% confidence level, 80% of all detectable events. This is based on a cost-benefit-analysis. Once this decision is made, the optimum frequency has been defined using the frequency distribution curve of recent data. Frequencies at or slightly above this optimum receive the highest points. Extra points

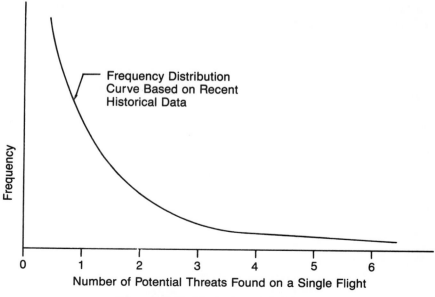

Figure 3-7. Typical patrol data.

should not be awarded for too much patrolling when the data does not support the benefit of higher frequencies.

An example point schedule is as follows:

Daily .... 15 pts
4 days per week .... 12 pts
3 days per week .... 10 pts
2 days per week .... 8 pts
Once per week .... 6 pts
Less than 4 times/month; more than once/month .... 4 pts
Less than once per month .... 2 pts
Never .... 0 pts

Select the point value corresponding to the actual patrol frequency.

This schedule is built for a corridor that has a frequency of third party intrusions that calls for an optimum patrol frequency of four days per week. In this case, the evaluator feels that daily patrols are perhaps justified and provide a measurably greater safety margin. Frequencies greater than once per day (once per 8 hour shift, for instance) warrant no more points than daily in this case.

The evaluator may wish to give point credits for patrols during activities such as close interval surveys (see *Corrosion Index*). Routine drive-bys, however, would need to be carefully evaluated for their effectiveness before credit is awarded.

CHAPTER FOUR

Corrosion Index

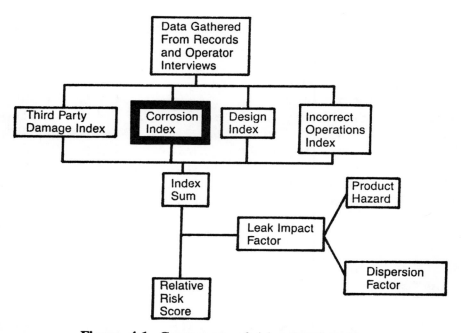

Figure 4-1. Components of risk rating flowchart.

Corrosion Risk

Corrosion Risk = (Atmospheric Corrosion)	20%
+ (Internal Corrosion)	20%
+ (Buried Metal Corrosion)	60%
	100%

Corrosion

The potential for pipeline failure caused either directly or indirectly by corrosion is perhaps the most familiar hazard associated with steel pipelines. This chapter will discuss common industry practices of corrosion analysis and mitigation. These practices will be incorporated into the risk management model. The complex mechanisms involved in the corrosion process will only be discussed to the extent required for an understanding of the industry practices.

Corrosion comes from the Latin word *corrodere,* meaning "gnaw to pieces." Corrosion, as it is used in this text, focuses mainly on a loss of metal from the pipe. From previous discussions of entropy and energy flow, we can look at corrosion from a somewhat esoteric viewpoint. Simply stated, manufactured metals have a natural tendency to revert to their original mineral form. While this is usually a very slow process, it does require the injection of energy to slow or halt the disintegration. Corrosion is of concern because any metal loss invariably means a reduction of structural integrity and hence an increase in risk of failure.

Non-steel pipeline materials are sometimes susceptible to other forms of environmental degradation. Sulfates and acids in the soil can deteriorate cement-containing materials such as concrete and asbestos-cement pipe. Some plastics degrade when exposed to ultraviolet light (sunlight). Polyethylene pipe can be vulnerable to hydrocarbons. Poly-vinyl chloride (PVC) pipe has been attacked by rodents that actually gnaw through the pipe wall. Most pipe materials can be internally degraded when transporting an incompatible product. All of these possibilities should be considered in this index. Even though the focus here is on steel lines, the evaluator can draw parallels to assess his non-steel lines in a similar fashion.

Some materials used in pipelines are not susceptible to corrosion and are virtually free from any kind of environmental degradation potential. These are not miracle materials by any means. Designers have usually traded away some mechanical properties such as strength and flexibility to obtain this property. Such pipelines obviously carry no risk of corrosion-induced failure. The *Corrosion Index* should reflect that lack of hazard.

The two factors that must be assessed are the material type and the environment. The environment includes the conditions that impact the pipe wall, internally as well as externally. Because most pipelines pass through several different environments, the assessment must allow for this either by sectioning appropriately or by considering each type of environment within a given section and using the worst case as the governing condition.

Several types of human errors can increase the risk from corrosion. Incorrect material selection for the environment (both internal and external exposures) is a common mistake. Placing incompatible materials close to each other can create or aggravate corrosion potentials. This includes joining materials such as bolts, gaskets, and weld metal. Welding processes must be carefully designed with corrosion potential in mind. These factors are covered in the *Incorrect Operations Index*.

In general, four ingredients are required for the commonly seen metallic corrosion to progress. There must exist an anode, a cathode, an electrical connection between the two, and an electrolyte. Removal of any one of these ingredients will halt the corrosion process. Corrosion prevention measures are designed to do just that.

The *Corrosion Index* consists of three categories: atmospheric corrosion, internal corrosion, and buried metal corrosion. This reflects three general environment types to which the pipe wall may be exposed.

Atmospheric corrosion deals with pipeline components that are exposed to the atmosphere. To assess the potential for corrosion here, the evaluator must look at items such as:

- susceptible facilities
    casings
    insulation
    'splash zone' locations
    supports, hangers
    ground/air interface
- atmospheric type
- painting/coating/inspection program

Atmospheric corrosion is weighted as 20% of the total corrosion risk in the example point schedules. The evaluator must determine if this is an appropriate weighting for his assessments.

Internal corrosion deals with the potential for corrosion originating within the pipeline. Assessment items include:

- product corrosivity
- preventive actions

Internal corrosion is also weighted as 20% of the total risk. Again, in certain situations, the evaluator may wish to give this category a different weighting.

Buried metal corrosion is the most complicated of the categories. Among the items considered in this assessment are:

- cathodic protection
- pipeline coatings

- soil corrosivity
- age of systems
- presence of other buried metal
- potential for stray currents
- stress corrosion cracking potential
- spacing of test leads
- inspections of rectifiers and interference bonds
- frequency of test lead readings
- close interval surveys
- use of internal inspection tools

Buried metal corrosion is weighted as 60% of the total corrosion risk picture. For non-metal lines, the evaluator must adjust this weighting to reflect the true hazards.

Especially in the case of buried metal, inspection for corrosion is done by indirect methods. Direct inspection of a pipe wall is often expensive and damaging (the coatings must be removed to see the pipe material). Corrosion assessments therefore usually infer corrosion potential by examining a few variables for evidence of corrosion. These inference assessments are then occasionally confirmed by direct inspection.

Because corrosion is often a highly localized phenomenon, and because indirect inspection provides only general information, uncertainty is great. With this difficulty in mind, many of the points of the *Corrosion Index* reflect the potential for corrosion to occur, which may or not mean that corrosion is actually taking place. Characteristics that may indicate a high corrosion potential are often difficult to quantify. For example, in buried metal corrosion, moist soil acts as the *electrolyte*—the environment that supports the electrochemical action necessary to cause this type of corrosion.

Electrolyte characteristics are of critical importance, but include highly variable items such as moisture content, aeration, bacteria content, ion concentrations. All these characteristics are location-specific and time-dependent. This makes them difficult to even estimate accurately. The parameters affecting atmospheric and internal corrosion potentials can be similarly difficult to estimate.

The evaluator should understand the limitations inherent in predicting corrosion. For this index, historical data can be of benefit. The evaluator can customize the item scoring to reflect more accurately the pipeline operator's experience with corrosion.

Items considered here reflect common industry practice in corrosion mitigation/prevention. The weightings indicate the relative importance of the item in terms of its contribution to the total corrosion risk.

A. Atmospheric Corrosion

In the United States alone, the estimated annual loss due to atmospheric corrosion is more than $2 billion [13]. Even though cross-country pipelines are mostly buried, they are not completely immune to this type of corrosion. Atmospheric corrosion is basically a chemical change in the pipe material resulting from the material's interaction with the atmosphere. Most commonly this interaction causes the oxidation of metal.

The evaluator may also include other types of degradations such as the effect of ultraviolet light on some plastic materials.

1. Facilities   **Suggested weighting 25%***
                                     Attribute (0–5 pts)

(*25% of atmospheric corrosion only)

The evaluator must determine the greatest risk from atmospheric corrosion by first locating the portions of the pipeline that are exposed to the most severe atmospheric conditions. Protection from this form of corrosion is considered in the next item. In this way, the situation is assessed in the most conservative manner. The most severe atmospheric conditions may be addressed by the best protective measures. However, the assessment will be the result of the worst conditions and the worst protective measures in the section. This conservatism not only helps in accounting for some unknowns, it also helps in pointing to situations where actions can be taken to improve the risk picture.

A schedule of descriptions of all the atmospheric exposure scenarios should be set up. The evaluator must decide which scenarios offer the most risk. This should be based on data (historical failures or discoveries of problems) when available, and employee knowledge and experience. The following is an example of such a schedule for steel pipe.

Air/water interface .... 0 pts
Casings .... 1 pts
Insulation .... 2 pts
Supports/hangers .... 2 pts
Ground/air interface .... 3 pts
Other exposures .... 4 pts
None .... 5 pts
Multiple occurrences detractor .... −1 pt

In this schedule, the worst case, the lowest point value, governs the entire section being evaluated.

Air/Water Interface. This is also known as a "splash zone," where the pipe is alternately exposed to water and air. This could be the result of wave or tide action, for instance. Sometimes called "waterline corrosion," the mechanism at work here is usually oxygen concentration cells. Differences in oxygen concentration set up anodic and cathodic regions on the metal. Under this scenario, the corrosion mechanism is enhanced as fresh oxygen is continuously brought to the corroding area and rust is carried away. If the water happens to be seawater or brackish (higher salt content), the electrolytic properties enhance corrosion as the higher ion content promotes the electrochemical corrosion process. Shoreline structures often illustrate the increased corrosion potential due to the air/water interface effect.

Casings. Industry experience points to buried casings as a prime location for corrosion to occur. Even though the casing and the enclosed carrier pipe are beneath the ground, atmospheric corrosion can be the prime corrosion mechanism. A vent pipe provides a path between the casing annular space and the atmosphere. In casings, the carrier pipe is often electrically connected to the casing pipe, despite efforts to prevent it. This occurs either through direct metallic contact or through a higher resistance connection such as water in the casing. When this connection is made, it is nearly impossible to control the direction of the electrochemical reaction, or even to know accurately what is happening in the casing. The worst situation occurs when the carrier pipeline becomes an anode to the casing pipe, meaning the carrier pipe loses metal as the casing pipe gains ions. Even without an electrical connection, the carrier pipe is subject to atmospheric corrosion, especially as the casing becomes filled with water and then later dries out (an air/water interface). The inability for direct observation or even reliable inference techniques causes this scenario to rate high in the risk hierarchy. (See Figure 4-2 and The Case For/Against Casings, page 69.)

Insulation. Insulation on aboveground pipe is notorious for trapping moisture against the pipe wall, allowing corrosion to proceed undetected. If the moisture is periodically replaced with fresh water, the oxygen supply is refreshed and corrosion is promoted. As with casings, such corrosion activity is usually not directly observable, and hence potentially more damaging.

Supports/Hangers. Another "hot spot" for corrosion by industry experience, pipe supports and hangers can often trap moisture against the pipe wall and sometimes provide a mechanism for loss of coating or paint. This occurs as the pipe expands and contracts, moving against the support and perhaps scraping away the coating. Mechanical-cor-

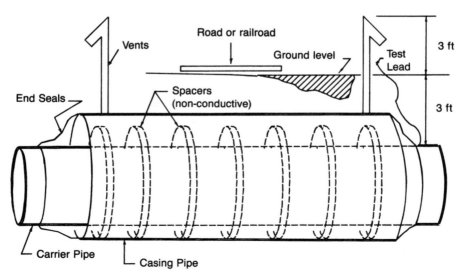

Figure 4-2. Typical casing installation.

rosion damage is also possible here. This type of damage often goes undetected.

Ground/Air Interface. As with the air/water interface, the ground/air interface can be harsh from a corrosion standpoint. This is the point at which the pipe enters and leaves the ground (or is lying on the ground). The harshness is caused in part by the potential for trapping moisture against the pipe (creating a water/air interface). Soil movements due to changing moisture content, freezing, etc., can also damage pipe coating, exposing bare metal to the electrolyte.

Other Exposures. The above cases should cover the range of worst-case exposures for steel pipe in contact with the atmosphere. One of the above situations must exist for aboveground piping; the pipe is either supported and/or it has one of the listed interfaces. A situation may exist however, where a non-steel pipe is not subject to degradation by any of the oxidation contributors listed. A plastic pipe may not be affected by any water or air or even chemical contact and yet may become brittle (and hence weaker) when exposed to sunlight. Sunlight exposure must therefore be included in that particular risk assessment.

None. If there is no corrodible portion of the pipeline exposed to the atmosphere, the potential for atmospheric corrosion does not exist.

Multiple Occurrences Detractor. In this example schedule, the evaluator deducts 1 point for sections that have multiple occurrences of a given condition. This reflects the increased opportunity for mishap because there are more potential corrosion sites. By this reasoning, a section containing many supports would receive $2 - 1 = 1$ pts, the equivalent of a section containing a casing. This says that the risk associated with multiple supports equals the risk associated with one casing. A further distinction could be made by specifying a point deduction for a given number of occurrences: -1 point for 5–10 supports, -2 points for 10–20 supports, etc. This may be an unnecessary complication, however.

Example:

A section of steel pipeline being evaluated has several road crossings in which the carrier pipe is encased in steel casing.

There are two aboveground valve stations in this section. One of these stations has approximately 25 ft of pipe supported on concrete and steel pedestals. The other one has no supports. The evaluator assesses the section for atmospheric corrosion "Facilities" as follows:

Casings .... *1 pts*
Ground/air interface .... *2 pts*
Supports .... *2 pts*

Picking the worst case, the point value for this section is 1 pt. The evaluator feels that the number of casings and number of supports and number of ground/air interfaces are roughly equivalent and chooses not to use the "multiple occurrences" option. If other sections being evaluated have a significantly different number of occurrences, adjustments would be needed to show the different risk picture. A distinction between a section with one casing and a section with two casings is needed to show the increased risk with two casings.

2. Atmospheric Type   **Suggested weighting 50%***
                                 Attribute (0–10 pts)

(*50% of atmospheric corrosion only)

Certain characteristics of the atmosphere can enhance or accelerate the corrosion of steel. They are thought to promote the oxidation process. Oxidation is the primary mechanism that is being evaluated in this section. Some of these atmospheric characteristics are:

- chemical composition
- humidity
- temperature

To make some simplifying generalities, we can say that:

- Chemical composition—Either naturally occurring airborne chemicals such as salt or CO_2, or man-made such as chlorine and SO_2 (which may form H_2SO_4 and H_2SO_3) can accelerate the oxidation of metal.

- Humidity—Because moisture can be a primary ingredient of the corrosion process, higher air moisture content is usually more corrosive.
- Temperature—Higher temperatures are more corrosive.

A schedule should be devised to show not only the effect of a characteristic, but also the interaction of one or more characteristics. For instance, a cool, dry climate is considered to be rather non-conducive to atmospheric corrosion. If a local industry produces certain airborne chemicals in this cool, dry climate however, the atmosphere might now be as severe as a tropical seaside location.

The following is an example schedule with categories for six different atmospheric types, ranked from most harsh to most benign, from a corrosion standpoint.

A. Chemical and marine .... 0 pts
B. Chemical and high humidity .... 2 pts
C. Marine, swamp, coastal .... 4 pts
D. High humidity, high temperature .... 6 pts
E. Chemical and low humidity .... 8 pts
F. Low humidity .... 10 pts

A. Chemical and Marine. Considered to be the most corrosive atmosphere, this includes certain offshore production facilities and refining operations in splash-zone environments. The pipe components are exposed to airborne chemicals and salt spray that promote oxidation, as well as routine submersion in water.

B. Chemical and High Humidity. Also a quite harsh environment, this may include chemical or refining operations in coastal regions. Airborne chemicals and a high moisture content in the air combine to enhance oxidation of the pipe steel.

C. Marine, Swamp, Coastal. High levels of salt and moisture combine to form a corrosive atmosphere here.

D. High Humidity and High Temperature. Similar to the situation above, this case may be seasonal or in some other way not as severe as condition C.

E. Chemical and Low Humidity. While oxidation-promoting chemicals are in the air, humidity is low, somewhat offsetting the effects. Distinctions may be added to account for temperatures here.

F. Low Humidity. The least corrosive atmosphere will have no airborne chemicals, low humidity, and low temperatures.

In applying this point schedule, the evaluator will probably need to use "comparables." The type of environment he is looking at might not specifically fit into one of these categories, but will usually be comparable to one of them.

Example:

The evaluator is comparing three atmospheric conditions. In the first case, he has a line that runs along a beach on Louisiana's Gulf Coast. This most closely resembles condition "C." Because there are several chemical producing plants nearby and winds may occasionally carry chemicals over the line, he adjusts the "C" score down by 1 pt to get 3 pts.

In the second case, he evaluates a steel line in eastern Colorado. While the line is seasonally exposed to higher temperatures and humidity, it is also frequently in cold, dry air. He assigns a point value based on condition "F" less 2 pts. This is 8 pts, equivalent from a risk standpoint to condition "E," even though there is no chemical risk.

Finally, he has a line in southern Arizona. Experience confirms that this environment does indeed support only minor corrosion. Because he foresees the evaluation of a line in a similarly dry, but also cold climate, he awards points for condition "F" −1 pt (for higher temperatures) = 9 points. (He plans to score the dry, cold climate as 10 pts.)

These evaluations therefore yield the following rank order and relative magnitude:

Louisiana .... *3 pts*
Colorado .... *8 pts*
Arizona .... *9 pts*

The evaluator sees little difference between conditions in Colorado and Arizona, from an atmospheric corrosion viewpoint, but feels that conditions around his line in south Louisiana are roughly three times worse.

3. Coating and Inspection

Suggested weighting 25%*
Prevention (0–5 pts)

(*25% of atmospheric corrosion only)

The third component of our study of the potential for atmospheric corrosion is an analysis of the preventive measures taken to minimize the potential. Obviously, where the environment is harsher, more preventive actions are required, and vice versa. From a risk standpoint, a situation where preventive actions are not required—a very benign environment—poses less risk than a situation where preventive actions are being taken to protect from a harsh environment.

The most common form of prevention for atmospheric corrosion is to isolate the metal from the offending environment. This is usually done with coatings. Coatings include paint, tape wraps, and a host of specially designed plastic coatings. For aboveground components, painting is by far the most common technique.

No coating is defect free, so the corrosion potential will never be totally removed, only reduced. How effectively the potential is reduced is dependent upon four factors:

- the quality of the coating
- the quality of the coating application
- the quality of the inspection program
- the quality of the defect correction program

Each of these components can be rated on a four point scale: good, fair, poor, or absent. The point values should probably be equivalent unless the evaluator can say that one component is of more importance than another. A quality coating is of little value if the application is poor; a good inspection program is incomplete if the defect correction program is poor. Perhaps an argument can be made that high scores in coating and application place less importance on inspection and defect correction. This would obviously be a sliding scale and is probably an unnecessary complication.

An evaluation scale could look like this:

good 3
fair 2
poor 1
absent 0

The evaluation values for each component will later be combined to get an overall rating for the item. To get the proper weighting for Coating/Inspection, we must convert the evaluation scale to a 5 point scale. Because the maximum evaluation score can be $4 \times 3 = 12$, we multiply the evaluation score by $5/12$ to report the Coating/Inspection value on a 5 point scale.

Remember, at this point, the evaluator is making no judgments as to whether a high quality coating or inspection program is needed. That determination is made when the attributes of facilities and atmosphere type are combined with an assessment of these preventions.

A. Coating. Evaluate the coating in terms of its appropriateness in its present application. Where possible, use data from coating stress tests to rate the quality. When this data is not available, draw from company experience.

Good—A high quality coating designed for its present environment.

Fair—An adequate coating but probably not specifically designed for its specific environment.

Poor—A coating in place but not suitable for long-term service in its present environment.

Absent—No coating present.

Note: Some of the more important coating properties include electrical resistance, adhesion, ease of application, flexibility, impact resistance, flow resistance (after curing), resistance to soil stresses, resistance to water, resistance to bacteria or other organism attack (in the case of submerged or partially submerged lines, marine life such as barnacles or borers must be considered).

B. Application. Evaluate the most recent coating application process and judge its quality in terms of attention to pre-cleaning, coating thickness, the application environment (temperature, humidity, dust, etc.), and the curing or setting process.

Good—Detailed specifications used, careful attention paid to all aspects of the application; appropriate quality control systems used.

Fair—Most likely a proper application, but without formal supervision or quality controls.

Poor—Careless, low quality application performed.

Absent—Application was incorrectly done, steps omitted, environment not controlled.

C. Inspection. Evaluate the inspection program for its thoroughness and timeliness. Documentation may also be an integral part of the best possible inspection program.

Good—Formal, thorough inspection performed specifically for evidence of atmospheric corrosion. Inspections are performed by trained individuals using checklists at appropriate intervals (as dictated by local corrosion potential).

Fair—Informal inspections, but performed routinely by qualified individuals.

Poor—Little inspection; reliance is on chance sighting of problem areas.

Absent—No inspection done.

Note: Typical coating faults include cracking, pinholes, impacts (from sharp objects), compressive loadings (stacking of coated pipes, for instance), disbondment, softening or flowing, and general deterioration (ultraviolet degradation, for example).

The inspector should pay special attention to sharp corners and difficult shapes. They are difficult to clean prior to painting, and difficult to adequately coat (paint will flow away from sharpness). Examples are nuts, bolts, threads, and some valve components. These are often the first areas to show corrosion and will give a first indication as to the quality of the paint job.

D. Correction of Defects. Evaluate the program of defect correction in terms of thoroughness and timeliness.

Good—Reported coating defects are immediately documented and scheduled for timely repair. Repairs are carried out per application specifications and are done on schedule.

Fair—Coating defects are informally reported and are repaired at convenience.

Poor—Coating defects are not consistently reported or repaired.

Absent—Little or no attention is paid to coating defects.

Example A:

In this section of aboveground piping, records indicate that a high quality paint was applied per NACE (National Association of Corrosion Engineers) specifications. The operator sends a trained inspector to all his aboveground sites once each quarter, and corrects all reported deficiencies at least twice per year. The evaluator awards points as follows:

Coating—good *3*
Application—good *3*
Inspection—good *3*
Defect correction—good *3*
                                                          12 points

Note: Twice per year defect correction is deemed appropriate for the section's environment.

This evaluation number is then converted to the point schedule (to carry the assigned weighting) by multiplying by 5/12:

12 × 5/12 = 5 points on the 5 point scale

Example B:

Here, a section contains several locations of aboveground pipe components at valve stations and compressor stations. Touch-up painting is done occasionally at the stations. This is done by a general contracting company at the request of the pipeline area foreman. No formal specifications exist. The foreman requests paint work whenever he feels it is needed (based upon his personal inspection of a facility). The evaluator awards points as follows:

Coating—fair *2.0*
Application—fair *1.8*
Inspection—fair *2.2*
Defect correction—poor *1.0*
                                                          7.0 points

Note: In this example, the evaluator wishes to make distinctions between the evaluation scores, so he uses decimals to rate items a little above or a little below the normal rating. This may be appropriate in some cases, but it adds a level of complexity that may not be warranted.

The evaluator feels that choice of paint is probably appropriate though not specified. Application is slightly below fair because no specifications exist and contractor work force is usually subject to regular turnovers. Inspection is slightly above fair because foreman does make specific inspections for evidence of atmospheric corrosion and is trained in spotting this evidence. Defect correction is poor because defect reporting and correction appear to be sporadic at best.

To correct for the 5 point scale:

$7.0 \times 5/12 = 2.9$ *points on a 5 point scale*

The Case For/Against Casings

Buried casings show up at several points in this risk assessment—sometimes as risk reducers, sometimes as risk creators. The following information provides a general discussion of the use of pipeline casings.

Casings. Oversized pipe, called casing pipe, is placed over the carrier pipeline to protect it from external loadings. Casings have long been used by the pipeline industry. They are generally placed under highways, roads, and railroads where higher external loadings are anticipated (Figure 4-2).

A casing also allows for easier replacement of the pipeline if a problem should develop. Instead of digging up a roadway, the pipeline can simply be pulled out of the casing, repaired, and reinstalled without disrupting traffic. (Unfortunately, this has proven to be a self-fulfilling prophecy. Failed pipelines have indeed been easily replaced because of the presence of casings, but it is the casings themselves that have often contributed to the failures!)

A third potential benefit from casings is that a slow pipeline leak can be contained in the casing and detected via the casing vent pipe rather than slowly undermining the roadway or forming underground pockets of accumulated product.

An industry controversy arises because the benefits casings provide are offset by problems caused by their presence. These problems are mostly corrosion-related. It is probably safe to say that corrosion engineers would rather not have casings in their systems. It is more difficult to protect an encased pipe from corrosion. The casing provides an environment where corrosion can proceed undetected. Because the pipeline cannot be directly inspected, indirect methods are used to give indications of corrosion. These techniques are not comprehensive, sometimes unreliable, and often require expert interpretation.

Some typical dilemmas/problems include:

Atmospheric corrosion can occur if any coating defects exist, and yet, insertion of the pipeline into the casing is an easy way to damage the coating and create defects.

End seals are used to keep water, mud, and other possible electrolytes out of the casing annular space. The presence of electrolyte in the annular space can lead to corrosion cells between the casing and the pipeline, as well as interference problems with the cathodic protection system. However, vent pipes are installed that allow direct communication between the casing annular space and the atmosphere—consequently, moisture is almost always present in the annular space.

Cathodic protection is usually employed to protect buried steel pipelines. The casing pipe can shield the pipeline from the protective currents if there is no electrical bond between the casing and the pipeline. If there is such a bond, the casing usually not only shields the pipeline from the current, but also draws current from it, effectively turning the pipeline into an anode that is sacrificed to protect the casing pipe which is now the cathode!

Several mitigative measures can be employed to reduce corrosion problems in casings. These are illustrated in Fig 4-2, and described below:

- Test leads. By comparing the pipe-to-soil potentials (voltages) of the pipeline versus the casing pipe, evidence of bonding between the two is sought. Test leads allow the voltage measurements to be made.
- Non-conductive spacers. These are designed to keep the pipeline physically (and hopefully electrically) separated from the casing pipe. They also help to protect the pipe coating during insertion into the casing.
- End seals. These are designed to keep the annular space free of substances that can act as an electrolyte (water, mud, etc.)
- Filling the annular space. Use of a dielectric (non-conductive) substance reduces the potential for electrical paths between the casing and the pipeline. Unfortunately, it also negates some of the casing benefits listed earlier.

Controversy shows up in this risk evaluation system also. Reflecting the tradeoff in benefits, casings can be risk reducers in the *Design Index* while also being risk adders in the *Corrosion Index* (atmospheric and buried metal). It would be nice to say that one will always outweigh the other, but we don't know that this is always the case.

From strictly a risk standpoint, casings can cause a maximum penalty of 9% in the *Corrosion Index* (using the sample point schedules in the text). This would be the situation for multiple casings in the section being evaluated, where no mitigative actions are being taken. A 9% offsetting benefit could be gained under the *Design Index* if the casings carry enough of the external loadings to create a pipe safety factor of about 30%. (Refer to the *Design Index* (Chapter 5) for a complete discussion of the pipe safety factor.) A risk cost/benefit analysis for casings can thus be performed.

Other factors must be considered in casing decisions. Often regulatory agencies leave no choice in the matter. The owner of the crossing (railroad, highway, etc.) may also mandate a certain design. Economics, of course, always play an important role. The costs of casings must include on-going maintenance costs, but the costs of not using casing must include pipe strong enough to carry all loads and damages to the crossing, should pipeline replacement be needed.

As an additional benefit of applying a risk management system such as this one to the problem of casings, the pipeline operator and designer have a rational basis for weighing the benefits of alternate designs.

B. Internal Corrosion

In this section, an assessment is made of the potential for internal corrosion. This is caused by a reaction between the inside pipe wall and the product being transported. Such corrosive activity may not be the result of the product INTENDED to be transported, but rather as the result of an impurity in the product stream. Seawater in an offshore natural gas stream, for example, is not uncommon. The methane will not harm steel, but salt water and other impurities can certainly promote corrosion of steel.

Microorganisms that can indirectly promote corrosion should also be considered here. Sulfate reducing bacteria and anaerobic acid producing bacteria are commonly found in oil and gas pipelines. They produce H_2S and acetic acid respectively, both of which can promote corrosion [27].

Pitting corrosion and crevice corrosion are specialized forms of galvanic or concentration cell corrosion commonly seen in cases of internal corrosion. Corrosion set up by an oxygen concentration cell can be accelerated if ions are present to play a role in the reactions. The attack against type 304 stainless steel by salt water is a classic example. Erosion as a form of internal corrosion is considered in the Mechanical-Corrosion Effects item of Buried Metal Corrosion.

Product reactions that do not harm the pipe material should not be included here. A good example of this is the buildup of paraffin in some oil lines. While such buildups cause operational problems, they do not normally contribute to the risk of pipeline failure unless they support or aggravate a corrosion mechanism that would otherwise not be present or as severe.

Some of the same measures used to prevent internal corrosion, such as internal coating, are used not only to protect the pipe, but also to protect the product from impurities that may be produced by corrosion. Jet fuels and high purity chemicals are examples of products carefully protected from such contaminants.

In a simple form, the assessment of risk due to internal corrosion need only examine the product characteristics and the preventive measures being taken to offset certain product characteristics.

1. Product Corrosivity

Suggested weighting 50%*
Attribute (0–10 pts)

(*50% of internal corrosion risk only)

The greatest risk exists in systems where the product is inherently incompatible with the pipe material. Next to this, the greatest risk occurs when corrosive impurities can routinely get into the product. A simple schedule can be devised to assign points to these product scenarios:

Strongly corrosive .... 0 pts
Mildly corrosive .... 3 pts
Corrosive only under special conditions .... 7 pts
Never corrosive .... 10 pts

"Strongly corrosive" suggests that a rapid, damaging kind of corrosion is possible. The product is highly incompatible with the pipe material. Transportation of brine solutions, water, products with H_2S, and many acidic products are examples of materials that are highly corrosive to steel lines.

"Mildly corrosive" suggests that damage to the pipe wall is possible but only at a slow rate. Having no knowledge of the product corrosivity can also fall into this category. It is conservative to assume that any product CAN do damage, unless we have evidence to the contrary.

"Corrosive only under special conditions" means that the product is normally benign, but there exists the chance of introducing a harmful component into the product. CO_2 or saltwater excursions in a methane pipeline is a common example. These natural components of methane production are usually removed before they can get into the pipeline. However, equipment used to remove such impurities is subject to equipment failures and subsequent spillage of impurities into the pipeline is a possibility.

"Never corrosive" means that there are no reasonable possibilities that the product transported will ever be incompatible with the pipe material.

The evaluator may interpolate and assign point values between the ones shown.

2. Internal Protection

Suggested weighting 50%*
Prevention (0–10 pts)

(*50% of internal corrosion risk only)

It is often economically advantageous to transport corrosive substances in pipe that is vulnerable to corrosion by the substance. In these cases, it is prudent to take actions to reduce or eliminate the damage. A point schedule, based upon the effectiveness of the action, will show how the risk picture is affected. In the following example schedule, points are added for each preventive action that is employed, up to a maximum of 10 points. Anti-corrosion activities being performed:

None	0 pts
Internal monitoring	2 pts
Inhibitor injection	4 pts

Internal coating .... 5 pts
Operational measures .... 3 pts
Pigging .... 3 pts

None. This, of course, means that no actions are taken to reduce the risk of internal corrosion.

Internal Monitoring. Normally, this is done in either of two ways: 1) by a probe that can continuously transmit electrical measurements that indicate a corrosion potential; or, 2) by a coupon that actually corrodes in the presence of the flowing product and is removed and measured periodically. Each of these methods requires an attachment to the pipeline to allow the probe or coupon to be inserted into and extracted from the flowing product.

Other methods include the use of a spool piece—a test piece of pipe that can be removed and carefully inspected for evidence of internal corrosion. Searching for corrosion products in pipeline filters or during pigging operations is another method of inspection/monitoring if done regularly.

To be creditable under this section, an inspection method requires a well-defined program of monitoring and interpretation of the data. It is implied that appropriate actions are taken, based upon the analysis from the monitoring program.

Inhibitor Injection. When the corrosion mechanism is fully understood, certain chemicals can be injected into the flowing product stream to reduce or inhibit the reaction. Because oxygen is a chief corroding agent of steel, an "oxygen scavenging" chemical can combine with the oxygen in the product to prevent this oxygen from reacting with the pipe wall. A more common kind of chemical inhibitor forms a protective barrier between the steel and the product—a coating, in effect. Inhibitor is reapplied periodically or continuously injected to replace the inhibitor that is absorbed or displaced by the product stream. In cases where microorganism activity is a problem, biocides can be added to the inhibitor. The evaluator should be confident that the inhibitor injection equipment is well-maintained and injects the proper amount of inhibitor at the proper rate. Inhibitor effectiveness is often verified by an internal monitoring program as described above.

A pigging program may be necessary to supplement inhibitor injection. The pigging would be designed to remove free liquids or bacteria colony protective coverings, which might otherwise interfere with inhibitor or biocide performance.

Internal Coating. New material technology allows for the creation of "lined pipe." This is usually a steel outer pipe that is isolated from a potentially damaging product by a material that is compatible with the product being transported. Plastics, rubbers, or ceramics are common isolating materials. They can be installed during initial pipe fabrication, during pipeline construction, or sometimes the material can be added to an existing pipeline.

Such two-material composite systems are discussed further in the *Design Index,* Chapter 5. For purposes of this section, the evaluator should assure himself that the composite system is effective in protecting the pipeline from damage due to internal corrosion. A common concern in such systems is the detection and repair of a leak that may occur in the liner before damage is done to the outer pipe.

The internal coating can be judged by the same criteria as coatings for protection from atmospheric corrosion and buried metal corrosion described in this chapter.

Operational Measures. In situations where the product is normally compatible with the pipe material but corrosive impurities can be introduced, operational measures are used to prevent the impurities. Systems used to dehydrate or filter a product stream fall into this classification. A system that strips sour gas (sulfur compounds) from a product stream is another example. Maintaining a certain temperature on a system in order to inhibit corrosion, would also be a valid operational measure. These systems or measures are termed *operational* here because the operation of the equipment is often as critical as the original design. Procedures and mechanical safeties should be in place to prevent corrosive materials from entering the pipeline in case of equipment failure or system overloads. The evaluator should check to see that the conditions for which the equipment was designed are still valid, especially if the effectiveness of the impurities removal cannot be directly determined. The evaluator should look for consistency and effectiveness in any operational measure purported to reduce internal corrosion potential.

Pigging. A pig is a cylindrical object designed to move through a pipeline for various purposes (Figure 4-3). Pigs are used to clean pipeline interiors (wire brushes are usually attached), separate products, push products (especially liquids), gather data (when fitted with special electronic devices), etc. A wide variety of special purpose pigs in many shapes and configurations is possible. There is even a by-pass pig that is designed with a relief valve to clear debris from in front of the pig if the debris causes a high differential pressure across the pig!

A regular program of running cleaning or displacement-type pigs to remove potentially corrosive materials is a proven effective method of reducing (but not eliminating) damage from internal corrosion. The program should be designed to remove liquids or other materials before they can do appreciable damage to the pipe wall. Monitoring of the materials displaced from the pipeline should include a search for corrosion products such as iron oxide in steel lines. This will help to assess the extent of corrosion in the line.

Pigging is partly an experience-driven technique. From a wide selection of pig types, the knowledgeable operator must choose an appropriate model, design the pigging mode including pig speed, distance, and driving force, and assess the progress during the operation. The evaluator should satisfy himself that the pigging operation is indeed beneficial and effective in removing corrosive products from the line in a timely fashion.

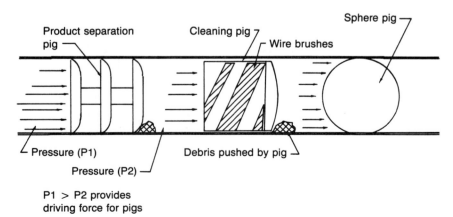

Figure 4-3. Examples of pipeline pigs.

Example:

A section of natural gas pipeline (steel) is being examined. The line transports gas from offshore production wells. The gas is dried and treated (removal of sulphur) offshore, but the offshore treating equipment malfunctions rather routinely. The operator injects inhibitor to control corrosion from any offshore liquids that escape the dehydration process. Recently, it was discovered that the inhibitor injector had failed for a period of two weeks before the malfunction was corrected. The operator also runs pigs once per month to remove any free standing liquids in the pipeline. Corrosion probes provide continuous data on the corrosion rate inside the line.

The evaluator assesses the situation as follows:

A. Product corrosivity .... *5 pts*
The line is exposed to corrosive components only under upset conditions, but 2 points are deducted because the upset conditions appear to be rather frequent.

B. Internal monitoring .... *2 pts*
    Inhibitor injection .... *2 pts*
    Operational measures .... *2 pts*
    Pigging .... *3 pts*
                                                                  ——
                                                      Total 9 pts

(10 pts max)

Points were deducted from each of two of the preventive measures because of known reliability problems with the actions. These two are the inhibitor injection and the operational measures. A penalty for the offshore operational measures was actually taken twice in this case, once in the product corrosivity and once in the preventive actions.

The total score for internal corrosion is then:

   A + B = 5 + 9 = 14 points.

C. Buried Metal Corrosion

This section will apply only to metallic pipe material that is buried in the earth and is subject to corrosion. If the pipeline being evaluated does not possess these characteristics, as would be the case for a plastic pipeline or a totally aboveground pipeline, the evaluator should use the previous two sections and any other pertinent factors to assess the corrosion risk.

Of the three categories of corrosion, this is the most complex. Several corrosion mechanisms can be at work in the case of buried metals. This situation is further complicated by the fact that corrosion activity is again normally deduced only from indirect evidence—direct observation is a rather limited option. The most common danger is from some form of galvanic corrosion.

Galvanic corrosion occurs when a metal or metals in an electrolyte (an electrically conductive fluid) form anodic and cathodic regions. A cathode is a metal region that has a greater affinity for electrons than the corresponding anodic region. This affinity for electrons is called the electronegativity. Different metals have different electronegativities and even different areas on a single piece of metal will have slightly different electronegativities. The greater the difference, the stronger the tendency for electrons to flow. If an electrical connection between anode and cathode exists, allowing this electron flow, metal will dissolve at the anode as metal ions are formed and migrate from the parent metal. Such a system, with anode, cathode, electrolyte, and electrical connection between anode and cathode, is called a galvanic cell and is illustrated in Figure 4-4.

Because soil is often an effective electrolyte, a galvanic cell can be established between a pipeline and another piece of buried metal, or even between two areas on the same pipeline. When a new piece of pipe is attached to an old piece, a galvanic cell can be established between the two metals. Dissimilar soils with differences in concentrations of ions, oxygen, or moisture can also set up anodic and cathodic regions on the pipe surface. Corrosion cells of this type are called "concentration cells." When these cells are established, the anodic region will experience active corrosion. The severity of this corrosion is dictated by variables such as the conductivity of the soil (electrolyte) and the relative electronegativities of the anode and cathode.

Common industry practice is to employ a two part defense against galvanic corrosion of a pipeline. The first line of defense is a coating over the pipeline. This is designed to isolate the metal from the electrolyte. If this coating is perfect, the galvanic cell is effectively stopped—the electric circuit is blocked because the electrolyte is no

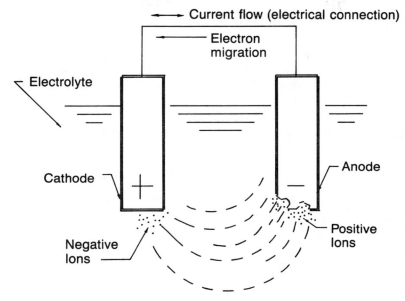

Figure 4-4. The galvanic corrosion cell.

longer in contact with the metal. It is safe to say, however, that no coating is perfect. If only at the microscopic level, defects will exist in any coating system.

The second line of defense is called cathodic protection. Through connections with other metals, the pipeline is turned into a cathode, which, according to the galvanic cell model, is not subject to loss of metal (as a matter of fact, the cathode actually gains metal). The theory behind cathodic protection is to ensure that the current flow is directed such that current flows TO the pipeline, and away from an installed bed of metal that is SUPPOSED to corrode. The installed metal that is to corrode is appropriately called sacrificial anode. The sacrificial anode has a lower affinity for electrons than the steel it is protecting. Depending on electrolyte (soil) type and some economic considerations, a voltage may be imposed on the system to further drive the current flow. When this is necessary, the system is referred to as an impressed current system (Figure 4-5).

In an impressed current system, rectifiers are used to drive the low-voltage current flow between the anode bed and the pipeline. The amount of current required is dictated by variables such as coating condition, soil type, anode bed design—all of which add resistance to this electric circuit.

There are many variables that influence the effectiveness of a corrosion control program for buried metal. In this evaluation, ten attributes and preventions are considered in assessing the potential for buried metal corrosion.

1. Cathodic Protection   Suggested weighting 13%*
                                       Prevention (0–8 pts)

(*13% of buried metal corrosion section only)

In most cases, some form of cathodic protection system is used to protect a buried steel pipeline. The exceptions might be instances where temporary lines are installed in fairly non-corrosive soil and where regulations do not require cathodic protection. Non-steel lines, of course, may not require corrosion protection.

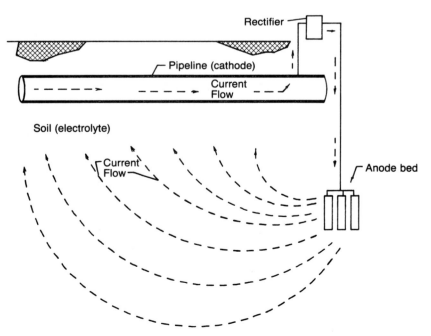

Figure 4-5. Pipeline cathodic protection with impressed current rectifier.

Points are given here for the existence of a system that meets the following general criteria:

- enough electromotive force is provided to effectively negate any corrosion potential.
- enough evidence is gathered, at appropriate times, to ensure that the system is working properly.

These criteria are expressed in general terms only at this point. More details of the maintenance and effectiveness of the cathodic protection system will be required in the next prevention items. Specifically, the presence and use of test leads and the use of close interval surveys are used to directly gauge the effectiveness of the cathodic protection system. These are examined in detail later.

By many regulatory agency requirements, a pipe-to-soil potential of at least -0.85 millivolts as measured by a copper-copper sulfate reference electrode, is the general cathodic protection level to meet the first criterion. The actual practice of ensuring adequate levels of cathodic protection is far more complex than this simple criterion. Readings must be carefully interpreted in light of the measurement system used. Too much current may damage the coating. Higher levels of protection are required when there is evidence of bacteria-promoted corrosion. A host of other factors must similarly be considered by the corrosion engineer in determining an adequate level of protection.

One aspect of the second criterion will be the maintenance of the associated cathodic protection equipment. For impressed current systems, equipment such as rectifiers must be maintained. Inspections of these pieces of equipment should perhaps be performed at shorter intervals than the overall check of the current levels. Because a rectifier provides the driving force for these cathodic protection systems, the operator must not allow a rectifier to be out of service for any length of time. Here the criterion of "appropriate times" should be used to assess the effectiveness of the system. Monthly or at least bimonthly rectifier inspections should be the norm.

A simple initial point schedule is recommended for this complex item.

General criteria are met...................................... 8 pts
General criteria are not met................................ 0 pts

To satisfy himself that the general criteria are being met, the evaluator should seek records of the initial cathodic protection design. Are the design parameters appropriate? What was the projected life span of the system? Is the system functioning according to plan?

The evaluator should then inspect documentation of the most recent checks on the system. Anode beds can become depleted, conditions can change, equipment can malfunction. Will the operator become aware of serious problems on the system in a timely manner? While cathodic protection problems should be caught during normal test lead checks and certainly during close interval surveys, problems such as broken rectifiers (or worse, rectifiers whose electrical connections have been reversed!) should be found even quicker.

Note: At this point in the evaluation, the level of protection over the entire pipeline is not so much in question as is the question of whether the operator has a system in place that CAN do the job, and will the operator quickly discover if the system quits working. The cathodic protection effectiveness over the entire pipeline is further assessed in following items.

Variations in the points given for this item should reflect the uncertainty of the evaluator. If he has no doubt that a properly engineered system was installed and is being prudently maintained, he should award 8 points. Doubts should bring the point level down. These doubts can be quantified as specific deductions for certain items, to ensure consistency.

2. Condition of Coating

Suggested weighting 17%*
Attribute (0–10 pts)

(*17% of buried metal corrosion only)

Pipeline coatings are often a composite of two or more layers of materials. Paints, plastics, and rubbers are common coating materials. A coating must be able to withstand a certain amount of mechanical damage from initial construction, from subsequent soil movements, and from temperature changes. The coating will be continuously exposed to ground moisture and any damaging substances contained in the soil. Additionally, the coating must adequately serve its main purpose: isolating the steel from the electrolyte. To do so, it must be fairly resistant to the passage of electricity. Because pipelines are designed for long life spans, the coating must perform all these functions without losing its properties over time—it must resist aging.

Typical coating systems include:

- cold-applied asphalt mastics
- layered extruded polyethylene
- fusion-bonded epoxy
- coal tar enamel and wrap
- tapes (hot or cold applied)

One of the main reasons for using cathodic protection systems is that no coating system is defect-free. Cathodic protection is designed to compensate for coating defects and deterioration. As such, one way to measure the condition of the coating is to measure how much cathodic protection is needed. Cathodic protection requirements are partially a function of soil conditions and the amount of exposed steel on the pipeline. Coatings with defects allow more steel to be exposed and hence require more cathodic protection. Cathodic protection is generally measured in terms of current consumption. A certain amount of voltage is thought to negate the corrosion effects, so the amount of current generated while maintaining this required voltage is a gauge of cathodic protection. A corrosion engineer can make some estimates of coating condition from these numbers.

One potentially bad situation that is difficult to detect is an area of disbonded coating, where the coating is separated from the steel surface. While the coating still provides a shield of sorts, moisture can often get between the coating and the steel. If this moisture is occasionally replaced, active local corrosion can proceed while showing little change in current requirements.

Another common type of coating defect is the presence of pinhole-size defects. These can be especially dangerous not only because they are difficult to detect, but also because they can promote narrow and deep corrosion pits. Because galvanic corrosion is an electrochemical reaction, a given driving force (voltage difference) will cause a set rate of metal ionization. If the exposed area of metal is large, the corrosion will be wide and shallow, whereas, a small exposure will lose the same volume of metal, causing deeper corrosion. Deeper corrosion is more weakening to the pipe wall because the wall strength is thickness-dependent. A small geometric discontinuity may also cause high stress concentrations (see Chapter 5).

To assess the present coating condition, several things should be considered, including the original installation process. An evaluation exactly like the one used to assess the coating for atmospheric corrosion protection is appropriate.

Again, no coating is defect free, therefore, the corrosion potential will never be totally removed, only reduced. How effectively the potential is reduced is dependent upon four factors:

- the quality of the coating
- the quality of the coating application
- the quality of the inspection program
- the quality of the defect correction program

Each of these components can be rated on a four-point scale: good, fair, poor, or absent. The weighting of each component should probably be equivalent unless the evaluator can say that one component is of more importance than another. A quality coating is of little value if the application is poor; a good inspection program is incomplete if the defect correction program is poor. Perhaps an argument can be made that high scores in coating and application place less importance on inspection and defect correction. This would obviously be a sliding scale and is probably an unnecessary complication.

An evaluation scale could look like this:

good	3
fair	2
poor	1
absent	0

Because a total of 12 points is possible, the coating score is multiplied by 10/12 to put the score on a 10-point scale. A 10-point scale is needed to keep the proper weighting (17% of the buried metal potential score) of this item.

A. Coating. Evaluate the coating in terms of its appropriateness in its present application. Where possible, use data from coating stress tests to rate the quality. Hardness, elasticity, adhesion to steel, and temperature sensitivity are common properties used to determine the appropriateness. When this data is not available, draw from company experience.

The evaluation should assess the coating's resistance to ALL anticipated stresses including a degree of abuse at initial installation, soil movements, chemical and moisture attack, temperature differentials, and gravity.

Good—A high quality coating designed for its present environment.

Fair—An adequate coating but probably not specifically designed for its specific environment.

Poor—A coating in place but not suitable for long-term service in its present environment.

Absent—No coating present.

Note: Some of the more important coating properties include electrical resistance, adhesion, ease of application, flexibility, impact resistance, flow resistance (after curing), resistance to soil stresses, resistance to water, resistance to bacteria or other organism attack (in the case of submerged lines, marine life such as barnacles or borers must be considered).

B. Application. Evaluate the most recent coating application process and judge its quality in terms of attention to pre-cleaning, coating thickness, the application environment (temperature, humidity, dust, etc.), and the curing or setting process.

Good—Detailed specifications used, careful attention paid to all aspects of the application; appropriate quality control systems used.

Fair—Most likely a proper application, but without formal supervision or quality controls.

Poor—Careless, low quality application performed.

Absent—Application was incorrectly done, steps omitted, environment not controlled.

C. Inspection. Evaluate the inspection program for its thoroughness and timeliness. Documentation will also be an integral part of the best possible inspection program. Inspection of underground coating can take several forms. Opportunities for visual inspection will occasionally present themselves, as the pipe is exposed for various reasons. When this happens, the operator should take advantage of the situation to have trained personnel evaluate the coating condition and record the findings.

A second inspection method, less direct than visual inspection, impresses a radio or electric signal onto the pipe and measures this signal strength at points along the pipeline (Figure 4-6). The signal strength should decrease linearly in direct proportion to the distance from the signal source. Peaks and unexpected changes in the signal indicate areas of non-uniform coating—perhaps damaged coating. This technique is called a holiday detection survey. Based upon the initial survey, test holes are dug for visual inspection of the coating in order to correlate actual coating condition with signal readings.

Another indirect method was mentioned in this section's introduction. A measure of the cathodic protection requirements—and especially the change in these requirements over time—gives an indication of the coating condition (Figure 4-8).

These methods discussed above and other indirect observation methods, require a degree of skill on the part of the operator and the analyzer. Industry opinion is divided on the effectiveness of some of these techniques. The evaluator should satisfy himself that the operator understands the technique and can demonstrate some success in its use for coating inspection.

Good—Formal, thorough inspection performed specifically for evidence of coating deterioration. Inspections are performed by trained individuals at appropriate intervals (as dictated by local corrosion potential). Full use of visual inspection opportunities in addition to one or more indirect techniques being used.

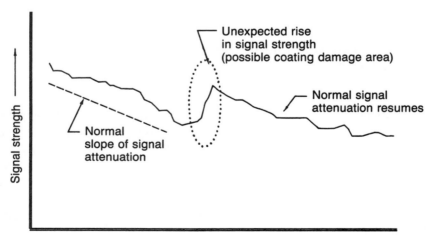

Distance from signal source (along pipeline)

Figure 4-6. Example of coating survey results.

Fair—Informal inspections, but performed routinely by qualified individuals. Perhaps an indirect technique is used but maybe not to its full potential.

Poor—Little inspection; reliance is on chance sighting of problem areas. Informal visual inspections when there is the opportunity.

Absent—No inspection done.

Note: Typical coating faults include cracking, pinholes, impacts (sharp objects), compressive loadings (stacking of coated pipes), disbondment, softening or flowing, and general deterioration (ultraviolet degradation, for example).

D. Correction of Defects. Evaluate the program of defect correction in terms of thoroughness and timeliness.

Good—Reported coating defects are immediately documented and scheduled for timely repair. Repairs are carried out per application specifications and are done on schedule.

Fair—Coating defects are informally reported and are repaired at convenience.

Poor—Coating defects are not consistently reported or repaired.

Absent—Little or no attention is paid to coating defects.

Example:

A buried oil pipeline in a dry, sandy soil is cathodically protected by sacrificial anodes attached to the line at a spacing of about 500 ft. A pipe-to-soil voltage measurement is taken twice each year over the whole section to ensure that cathodic protection is adequate. Records indicate that the line was initially coated with a polyethylene material that was extruded over the sand-blasted and primed pipe. An inspector supervised the coating process. The pipe-to-ground potential has not changed measurably since original installation. This section of line has not been exposed for ten years.

The evaluator assess the situation as follows:

Cathodic protection *8pts*
Coating condition
    coating (good) .... *3 pts*
    application (good) .... *3 pts*

inspection (fair) .... *2 pts*
defect correction (good) .... $\underline{3\ pts}$
$11\ pts$
(correction for 10 point scale: 11 × 10/12 = 9.2) $\underline{9\ pts}$
(Cathodic Protection) + (Coating Condition) = 8 + 9 = 17 pts

For inspection, the evaluator feels that the semi-annual pipe-to-soil voltage readings give a good indication of coating condition. Full points would be awarded if this was confirmed by visual inspection (cracked or disbonded coating may not be found by the potential readings alone). Defect correction is an unknown at this point. Three points are awarded based upon the thoroughness with which the operator runs other aspects of his operations—in other words, some benefit of the doubt is given here. Coating selection and application processes appear to be high quality, based upon records and conversations with the operator.

The coating condition assessment can be made more data-driven if accurate measurements of cathodic protection current requirements exist. These measurements are usually in the form of milli-amperes per square foot of pipeline surface area. A schedule such as the following could be used:

Current Requirements	Coating Condition
0.0003 mA/sq ft	good
0.003 mA/sq ft	fair
0.1 mA/sq ft	poor
1.0 mA/sq ft	absent

Again, less current requirements mean less exposed metal and better electrical isolation from the electrolyte.

Coating condition is considered to be an attribute in this evaluation system. It is an element of the risk picture that is not easy to change.

3. Soil Corrosivity **Suggested weighting 7%***
Attribute (0–4pts)

(*7% of buried metal corrosion only)

Because the coating system should be considered to be an imperfect barrier, the soil will necessarily contact the pipe wall. Soil corrosivity is primarily a measure of how well the soil can act as an electrolyte to promote galvanic corrosion on the pipe. Secondarily, elements of the soil that may directly or indirectly promote corrosion mechanisms should also be considered. These include bacterial activity and the presence of corrosive substances.

The effect of the soil condition on the coating itself is NOT a part of this item. Soil effects on the coating itself (mechanical damage, moisture damage, etc.) should be considered when judging the coating selection in item B.

The importance of soil as a factor in the galvanic cell activity is not a widely agreed upon subject. Historically, the soil's resistance to electrical flow was the measure used to judge the contribution of soil effects to galvanic corrosion. As with any component of the galvanic cell, the electrical resistances play a role in the operation of the circuit. Soil resistivity is dependent upon variables such as moisture content, porosity, temperature, and soil type. Some of these are time dependent or seasonal variables, corresponding to rainfall or atmospheric temperatures. The evaluator may wish to change the weighting of this attribute if he feels that it plays a greater (or lesser) role in buried metal corrosion potential.

Microorganism activity can promote corrosion. A family of anaerobic bacteria (no oxygen needed for the bacteria to reproduce), called sulfate-reducing bacteria, can cause the depletion of the hydrogen layer adjacent to the outside pipe wall. This hydrogen layer normally provides a degree of protection from corrosion. As it is removed, corrosion reactions can actually be accelerated. Soils with sulfates or soluble salts are favorable environments for anaerobic sulfate-reducing bacteria [22].

While it doesn't actually attack the metal, the microorganism activity tends to produce conditions that accelerate corrosion. The sulfate reducing bacteria are commonly found in areas where stagnant water or water-logged soil is in contact with the steel. Upon excavation, evidence of bacterial activity is sometimes seen as a layer of black iron-sulfide on the pipe wall. An oxidation-reduction probe can be used to test for conditions favorable for bacteria activity (does not determine if corrosion is taking place, however). A normal cure for microorganism-promoted corrosion is increased levels of cathodic protection currents.

Different pipe materials are susceptible to damage by various soil conditions. Sulfates and acids in the soil can deteriorate cement-containing materials such as concrete or asbestos-cement pipe. Polyethylene pipe may be vulnerable to damage by hydrocarbons. Special information of pipe material susceptibility to soil components should be incorporated into this section.

For metals, more acidic (lower pH) soils promote corrosion more than the more alkaline (higher pH) soils. The soil pH may similarly affect other pipe materials. Therefore, pH can also be a factor in this analysis.

The general soil conductivity is dependent upon factors such as moisture content, ion concentrations, and soil components. A schedule can be developed to assess the average or worst case (either could be appropriate—choice must be consistent across all sections evaluated, though) soil resistivity. This is a broad-brush measure of the electrolytic characteristic of the soil.

Low resistivity (high corrosion potential)
   < 500 ohm-cm of soil.................................... 0 pts
Medium 500–10,000 ohm-cm .... 2 pts
High resistivity (low corrosion potential)
   > 10,000 ohm-cm .... 4 pts
Don't know .... 0 pts
Special situation .... −1 to −4 pts

A special situation such as evidence of high microorganism activity or unusually low pH that promotes steel oxidation, should be accounted for by reducing the point value (but not below zero points). Not knowing the soil corrosion potential would conservatively warrant a score of zero points.

4. Age of System   **Suggested weighting 5%***
                                      Attribute (0–3 pts)

(*5% of buried metal corrosion only)

Most pipeline systems are designed for a useful service life of 30 to 50 years. Some have been in service much longer. Years in service alone, then, is not a reliable indicator of pipeline risk. On the other hand, more years in service increases the area of opportunity for something to go wrong. A risk assessment would be incomplete without addressing the age issue, but the amount of importance placed on this item is arguable.

Because age itself is not a failure mechanism, age is included as a variable portion—a contributing factor—of one of the failure modes. There are theories of metallurgical changes in pipe materials that may only have an appreciable effect after years of burial or impressed current exposure. Of the four indexes in this risk evaluation model, the age variable could logically fit as a contributing factor in either the *Design Index* (a factor in fatigue loadings) or here in the *Corrosion Index* (time is a factor in all forms of corrosion). It has been included here.

The cut off between age groups is arbitrary. It is not thought that a system suddenly makes large jumps in risk exposure at its start-up date anniversary each year. Divisions smaller than one year will be an unnecessary complication in most cases, but the option of scaling by month, week, or even day, exists, of course.

A point schedule such as:

0 to 5 years in service .... 3 pts
5 to 10 .... 2 pts
10 to 20 .... 1 pt
more than 20 years in service .... 0 pts

implies that younger pipelines have less risk exposure (all other factors being equal) up to 20 years in service. Beyond that, no credit is given for fewer years.

Example:

A line that has been in service for 11 years would be assessed:
    1 point
A line that has been in service for 3 years would be assessed:
    3 points

5. Current Flow to   **Suggested weighting 7%***
   Other Buried Metal   **Attribute (0–4 pts)**

*(*7% of buried metal corrosion only)*

The presence of other buried metal in the vicinity of a buried metal pipeline is a potential source of risk. Other buried metal can short circuit or otherwise interfere with the cathodic protection system of the pipeline. Even in the absence of a cathodic protection system, other metal can establish a galvanic corrosion cell with the pipeline. This may cause corrosion on the pipeline. The common term for these effects is interference.

The most critical interference situations, which should not be tolerated even for short periods, occur when there is physical contact between the pipeline and the other metal. This is especially critical when the other metal has its own impressed current system. Electric railroads are a good example of systems that can cause special problems for pipelines whether or not physical contact occurs. The danger occurs when the other system is competing with the pipeline for electrons. If the other system has a stronger electronegativity, the pipeline will become an anode, and depending upon the difference in electron affinity, the pipeline can experience accelerated corrosion. As mentioned earlier, coatings may actually worsen the situation if all the anodic metal dissolves from pinhole-size areas, causing narrow and deep corrosion pits.

Common mitigation measures for interference problems include interference bonds, isolators, and test leads. Interference bonds are direct electrical connections that allow the **controlled** flow of current from one system to another. By controlling this flow, corrosion effects arising from the foreign systems can be controlled. Isolators, when properly installed, can similarly control the flow of current. Finally, test leads are used to monitor for problems. By comparing the pipe-to-soil potential readings of the two systems, signs of interference can sometimes be found. As with any monitoring system, test leads must be used regularly by trained personnel, and corrective actions must be made when problems are identified.

A reasonable question now is *How close is close?* The proximity of the foreign metal obviously is a key factor in the risk potential, but the distance is not strictly measured in feet or meters. Longer distances can be dangerous in low resistivity soil or in cases where the current levels are relatively high. A reasonable rule of thumb might be to consider all buried metal within 500 ft of the pipeline. This rule should be tailored to the specific situation, but then held constant for all pipelines evaluated.

Points should be assessed based upon how many occurrences of buried metal exist along a section. Again, the greater the area of opportunity, the greater the risk. For pipelines in corridors with foreign pipelines, higher levels of risk exist.

Because almost any situation is potentially hazardous, a distinction based on severity of the situation might get quite complex. The following example schedule gives equal weighting to all situations: parallel pipelines, crossing pipelines, casings, buried insulating flanges, etc. Credit for mitigation measures can be given.

Number of Occurrences	Points
none	4
1–10	2
11–25	1
>25	0

If, in every instance of an occurrence, prevention/mitigation measures are taken and monitored for effectiveness, double the point value up to a maximum of 3 points. By doing this, prevention measures reduce the risk, but never to the extent of not having any potentially hazardous situations present.

Example:

In this section of pipeline (steel), the evaluator finds six cased road crossings, three crossings of foreign pipelines, and two instances of parallel pipelines within 200 ft of the water pipeline. Each road casing has test leads attached to detect possible short circuits (physical contact or low resistivity contact). The foreign pipeline crossings are each connected to the water line by interference bonds that are monitored regularly. Annual close interval surveys are specifically designed to monitor the areas of parallel pipelines.

The evaluator records points as follows:

6 + 3 + 2 = 11 instances ....*1 pt*

Credit is given for mitigation because all instances are addressed to the satisfaction of the evaluator.

Final score=1 pt × 2 = 2 pts

6. AC Interference　　　　　　**Suggested weighting 7%***
　　　　　　　　　　　　　　　　Attribute (0–4pts)

(*7% of buried metal corrosion only)

Pipelines near to AC power transmission facilities are exposed to a unique risk. Through either a ground fault or a process known as induction, the pipeline may become electrically charged. Not only is this charge potentially dangerous to people coming into contact with the line, it is also potentially dangerous to the pipeline itself. Electric current seeks the path of least resistance. A buried steel conduit like a pipeline may be an ideal path for some distance. Almost always, though, the current will eventually jump from the pipeline to another more attractive path. The locations where the current enters or leaves the pipe may cause severe metal loss as the electrical charge arcs to or from the line. At a minimum, the pipeline coating may be damaged by the AC interference effects.

The ground fault scenario of charging the pipeline includes the phenomena of conduction, resistive coupling, and electrolytic coupling. It can occur as AC power travels through the ground from a fallen transmission line, an accidental electrical connection onto a tower leg, through a lightening strike on the power system, or from an imbalance in a grounded power system. These are often the more acute cases of AC interference, but they are also often the more easily detectable cases. The sometimes high potentials resulting from ground faults expose the pipe coating to high stress levels. This occurs as the soil surrounding the pipeline becomes charged, setting up a high voltage differential across the coating. Disbondment or arcing may occur. If the potentials are great enough, the arcing may damage the pipe steel itself.

The induction scenario occurs as the pipeline is affected by either the electrical or magnetic field created by the AC power transmission. This sets up a current flow or a potential gradient in the pipeline (Figure 4-7). These cases of capacitive or inductive coupling are dependent upon such factors as the geometrical relation of the pipeline to the power transmission line, the magnitude of the power current flow, the frequency of the power system, the coating resistivity, the soil resistivity, and the longitudinal resistivity of the steel [26].

Formulae exist to estimate the potential effects of AC interference under normal and fault conditions. To perform these calculations, some knowledge of power transmission load characteristics of the power system is required. Estimations and measurements will be needed to generate soil, coating, and steel resistivity values, as well as the distances between the pipeline and the power transmission facilities. The key factors in assessing the normal effects for most situations will most likely be the characteristics of the AC power and the distance from the pipeline. Fault conditions can, of course, encompass a multitude of possibilities.

Methods used to minimize the AC interference effects, both to protect the pipeline and/or personnel coming into contact with the line include [20]:

- electrical shields
- grounding mats
- independent structure grounds

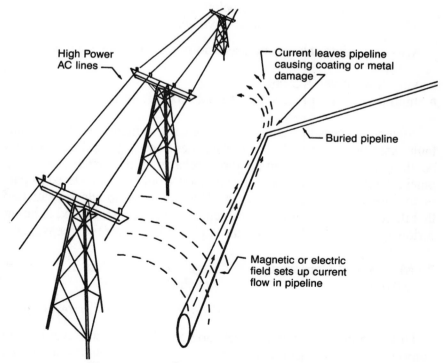

Figure 4-7. AC power currents on pipeline.

- bonding to existing structures
- distributed anodes
- casings
- proper use of connectors and conductors
- insulating joints
- electrolytic grounding cells
- polarization cells
- lightning arresters

Monitoring should be an integral part of the AC mitigation effort.

Because there are so many variables involved in performing accurate calculations, a simplified schedule is recommended for this rather complex issue. In terms of risk exposure, one of three possible scenarios can be said to exist:

No AC power is within 500 ft of the pipeline .... 4 pts
AC power is nearby, but preventive measures are
    being used to protect the pipeline .... 2 pts
AC power is nearby, no preventive actions taken .... 0 pts

Also fitting into the second scenario might be cases such as:

- very low power AC only
- high power AC present but at least 1,000 ft away

Preventive measures can be designed for induction or for ground fault cases or for both. As previously mentioned, grounding cells can be designed to safely handle the discharging of current from the pipeline. Close monitoring of the situation would be considered as part of the preventive measures taken. The evaluator should be satisfied that the potential AC current problem is well understood and is being seriously addressed, before credit is given for preventive measures.

7. Mechanical Corrosion Effects   Suggested weighting .... 8%*   Attribute (0–5 pts)

(*8% of buried metal corrosion only)

This item includes damaging phenomena that consist of a corrosion component and a mechanical component. This includes hydrogen stress corrosion cracking (HSCC), sulfide stress corrosion cracking

(SSCC), hydrogen induced cracking (HIC) or hydrogen embrittlement, corrosion fatigue, and erosion.

Stress corrosion cracking (SCC) can occur under certain combinations of physical and corrosive stresses. It is characterized by formation of corrosion-accelerated cracking in areas of the pipe wall subjected to high stress levels. The presence of corrosive substances aggravates the situation. Certain types of steel are more susceptible than others. In general, a steel with a higher carbon content is more prone to stress-corrosion cracking. Characteristics of the steel that may have been brought about by welding or other post-manufacturing processes, may also make the steel more susceptible. Materials that have little fracture toughness (see *Design Index*) do not offer much resistance to brittle failure. Rapid crack propagation brought on by corrosion and stress is most likely in these materials.

Stress corrosion cracking is difficult to detect. SCC failures are not predictable. Even a fairly non-corrosive environment can contribute greatly to a SCC process. The effects can be highly localized. A previous history of this type of process may be the best evidence of susceptibility.

In the absence of historical data, the susceptibility of a pipeline to this sometimes violent failure mechanism should be judged by identifying conditions that may promote the SCC process.

Contributing factors:

Stress. Tensile stress at the pipe surface is thought to be a necessary condition. The stress might be residual, however, and hence virtually undetectable. The higher the stress, the more potential for crack formation and growth. It is reasonable to assume that all pipelines will be under at least some amount of stress. Because internal pressure is often the largest stress contributor, pipelines operating at higher pressures are thought to have more susceptibility to SCC. Thermally induced stresses and bending stresses can also contribute to the overall stress level, but, for simplicity sake, only internal pressure is used as a factor in assessing potential for SCC.

Environment. High pH levels close to the steel can contribute. This may mean a high pH in the soil, in the product, or even in the coating. Chlorides, H_2S, CO_2, and high temperatures are more contributing factors. The presence of certain bacteria will increase the risk. In general, any environmental characteristics that promote corrosion should be considered to be risk contributors here. This must include external and internal contributors.

Steel type. As already stated, a high carbon content (>.28%) increases the likelihood of stress corrosion cracking. Low ductility materials with low fracture toughness are more susceptible. Sometimes the rate of loading determines the fracture toughness—a material may be able to withstand a slow application of stress, but not a rapid application (see *Design Index*). This further complicates the use of material type as a contributing factor.

A schedule can be developed that employs the first two contributing factors in an assessment of the potential for SCC. Table 4-1 shows stress level is related to the pipeline environment. Low stress in a benign environment is the best condition (lower left part of table), while high stress in a corrosive environment is the most dangerous condition (upper right part of table). Stress level is expressed as a percentage of Maximum Allowable Operating Pressure (MAOP)—the highest normal operating pressure divided by MAOP. The environment is scored by adding the *product characteristic score* (taken from *Internal Corrosion*, this is a 0 to 10 point scale, p. 71) to the *soil corrosivity* score (a 0 to 4 point scale, p. 88).

Note: If the section being evaluated is an aboveground section, use the *atmospheric type* (p. 62) score instead of the *soil corrosivity* score. Scale the *atmospheric type* score down to a 0 to 4 point scale, in order for it to have the same relative impact as the *soil corrosivity*.

Table 4-1

% MAOP:	0–20%	21%–50%	51%–75%	>75%
Environment*				
0	3	2	1	1
4	4	3	2	1
9	4	4	3	2
14	5	5	4	3

* *Environment = (product corrosivity) + (soil corrosivity)*
minimum = 0 points; maximum = 14 points

History of stress corrosion cracking should be seen as the strongest evidence of this risk and should accordingly score the section at 0 points.

Example:

The evaluator is assessing a natural gas pipeline that is rated for an MAOP of 1,500 psig. The line never exceeds 800 psig in the section being evaluated. The natural gas is sometimes corrosive and received a score of 4 points when it was evaluated for internal corrosion. The soil corrosivity score is 3 points for this normally dry, sandy soil.

From the table, the evaluator uses an environmental number of $4 + 3 = 7$, and a stress number of $800/1,500 = 53.3\%$ to get a point value of 3. He has rounded the number up to 3 because the stress level is so close to the cutoff 51% number.

Erosion. This is the removal of pipe wall material caused by the abrasive or scouring effects of substances moving against the pipe wall. It is a form of corrosion only in the pure definition of the word, but is considered here as a mechanical corrosion effect.

High velocities and abrasive particles in the product stream are the normal contributing factors. Impingement points such as elbows and valves are the most susceptible erosion points. Gas at high velocities may be carrying entrained particles of sand or other solid residues and, consequently, can be especially damaging to the pipe components.

Historical evidence of erosion damage is a strong indicator of susceptibility. Other evidence includes high product stream velocities (look for large pressure changes in short distances), or abrasive fluids. Combinations of these factors are, of course, the strongest evidence.

Point values for mechanical corrosion effects should be reduced by 2 when factors are right for erosion damage to occur. If, in the above example, the evaluator is told that sand is sometimes found in filters or damaged valve seats, and that some valves had to be replaced recently with more abrasion-resistant seat materials, he would have sufficient reason to deduct 2 points for this item.

Erosion factor .... −2 points

8. Test Leads   **Suggested weighting 10%***
                                      Prevention (0–6 pts)

(*10% of buried metal corrosion only)

Perhaps the primary method for monitoring the effectiveness of a cathodic protection system is through the use of test leads, wires attached (normally welded or soldered) to the buried pipeline and extended above ground. A test lead allows a trained technician to attach a voltmeter with a reference electrode and measure the pipe-to-soil potential. Such a measurement indicates the degree of cathodic protection on the pipe because it is indicating the tendency of current flow, both in terms of magnitude and direction (to the pipe, or from the pipe) (Figure 4-5).

In the interpretation of these measurements, attention must be paid to the resistances that are part of the pipe-to-soil reading. The reading which is sought, but difficult to obtain, is the electric potential difference between the outside surface of the pipe and a point in the adjacent soil a short distance away. In actual practice, a reading is taken between the pipe surface (via the test lead) and a point at the ground surface, usually several feet above the pipe. The circuit is completed at the ground surface by contacting the soil with a reference electrode (a half cell, usually copper electrode in a copper sulfate solution). This reading measures not only the piece of information sought, but also all resistances in the electric circuit, including wires, pipe steel, instruments, connectors, and, the largest component, the several feet of soil between the buried pipe wall and the ground surface. The knowledgeable corrosion engineer will take his readings in such a way as to enable him to separate the extraneous information from the data he needs. The industry refers to this technique as compensating for the IR drop.

Chemical reactions occur at the anode and the cathode as ions are formed. The soil component of the circuit is a nonmetallic current path. Consequently, this model is not directly analogous to a simple electrical circuit. Of primary interest to the corrosion engineer is a measure of the cathodic protection effectiveness. There is some controversy in the industry as to exactly how the readings should be interpreted in terms of the IR drop. In many cases the controversy is theoretical only because government regulations mandate certain techniques. The evaluator should be satisfied himself that sufficient expertise exists in the interpretation of readings to give valid answers.

Placement of test leads at locations where interference is possible is especially important. The most common points are metal pipe casings and foreign pipelines. At these sites, careful attention should be paid to the direction of current flow to ensure that the pipeline is not anodic to the other metal. Where pipelines cross, test leads on both lines can show if the cathodic protection systems are competing.

Because galvanic corrosion can be a localized phenomenon, the test leads are only indicators of cathodic protection in the immediate area around the lead. Closer test lead spacing, therefore, yields more information and less chance of large areas of active corrosion going undetected. Because corrosion is a time-dependent process, the number of times the test leads are monitored is also important.

Using these concepts, a point schedule is developed as follows:

All buried metal in the vicinity of the pipeline is monitored directly by test leads, and test lead spacing is no greater than one mile throughout this section .... 3 pts

Test leads are spaced at distances 1 to 2 miles apart (maximum) and all foreign pipeline crossings are monitored via test leads. Not all casings are monitored; there may be other buried metal which is not monitored .... 1–2 pts

Test lead spacing is sometimes more than 2 miles; not all potential interference sources are monitored .... 0 pts

Frequency of readings at test leads:
Pipe-to-soil readings are taken with the IR drop understood and compensated, at intervals of:
<6 months .... 3 pts
6 months–annually .... 2 pts
>annually .... 1 pt
Notes: As previously explained, lack of proper IR drop compensation may negate the effectiveness of all the readings.

Readings taken at intervals of greater than one year do have some value, but a year's worth of corrosion might have proceeded undetected between readings.

Add points for spacing to the points for the frequency of readings, maximum of 6 points, minimum of 0 points.

Example:

A section of gas line is being evaluated here. In this section, test leads are spaced closer than 2 miles apart, but there are two foreign pipeline crossings that are not monitored. Pipe-to-soil readings are taken every year. Points are awarded as follows:

Spacing <2 miles, but not every crossing
   monitored .... *1 pt*
Readings every year ....*2 pts*
                                                          _____
                                                          3 pts

9. Close Interval Surveys
Suggested weighting 13%*
Attribute (0–8 pts)

(*13% of buried metal corrosion only)

A powerful tool in the corrosion engineer's toolbag is a variation on test lead monitoring called close interval surveying. In this technique, pipe-to-soil readings are taken and IR compensation is employed, only now the readings are taken every 2 feet to 15 feet along the entire length of the pipeline. In this way, almost all localized interference or potential corrosion activity can be detected.

Any aboveground pipeline attachment, including valves, test leads, casing vents, etc. can be used to connect to one side of a voltmeter. The other side of the voltmeter is connected by a wire to the reference half cell that is used to make electrical connection at the ground surface as the surveyor walks along the pipeline. The voltmeter and data logging device are therefore in the circuit between the two electrodes. Results are usually interpreted from a strip chart that will show peaks and valleys when the current flow changes magnitude or direction. (See Figure 4-8).

Ideally such a profile of the pipe-to-soil potential readings will indicate areas of interference with other pipelines, casings, etc., areas of inadequate cathodic protection, and even areas of bad coating. Often, excavations are performed to verify the survey readings. A close interval survey should be done periodically to pick up changes along the pipeline route. The survey's role in risk reduction is quantified in the following point equation:

Minimum Requirements
A thorough close interval survey has been performed over the entire pipeline section by trained personnel. Interpretations of all readings were made by a knowledgeable corrosion engineer.

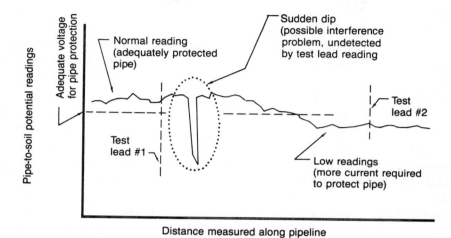

Figure 4-8. Close interval pipe-to-soil potential survey.

Corrective actions based upon survey results have been taken or are planned (in a timely fashion).

Timeliness

8 − (years since survey) = point value

Example:

A survey which met all the requirements was performed 3 years ago. Points awarded are:

8 − 3 = 5 points

Point values for this prevention are higher than most others in this section. This reflects the fact that this particular technique is quite robust in monitoring the condition of buried steel pipelines. It is also a proactive technique—detecting potential problems hopefully before appreciable damage is done to the pipeline.

10. Internal Inspection Tool   Suggested weighting 13%*
**   Prevention (0–8 pts)**

(*13% of buried metal corrosion only)

The use of instrumented pigs to inspect a pipeline from the inside is a rapidly maturing technology. While available for nearly 30 years, this technique is presently benefiting from advancements in electronics and computing technology making it much more useful to the pipeline industry. Any change in pipe wall thickness can theoretically be detected. These devices can also detect pipe wall cracks, laminations, and other material defects. Coating defects may someday also be detected in this fashion.

The most common "intelligent pigs" employ either an ultrasonic or a magnetic flux technology to perform the inspection. The ultrasonic devices use sound waves to continuously measure the wall thickness around the entire circumference of the pipe as the pig travels down the line. The thickness measurement is obtained by measuring the difference in travel time between sound pulses reflected from the inner pipe wall and the outer pipe wall. A liquid couplant is required to transmit the ultrasonic waves from the transducer to the pipe wall. This makes the device difficult to use in gas lines that must be kept dry.

The magnetic flux pig sets up a magnetic field in the pipe wall and then measures this field. Changes in the pipe wall will change the magnetic field. This device emphasizes the detection of anomalies rather than measurement of wall thickness, although experienced personnel can closely estimate defect sizes and wall thicknesses.

In either case, all data is recorded. Both types of pigs are composed of several sections to accommodate the measuring instruments, the recording instruments, a power supply, and cups used for propulsion of the pig.

Where the intelligent pigs have been used, favorable results have been reported. Many of the other indicators of corrosion are indirect and are masked by disbonded coating or can be averaged out because corrosion is often a localized phenomenon. Internal inspection devices are much more direct indicators of corrosion activity. As the technology matures, this will no doubt be an integral part of every pipeline monitoring program. Because this technique discovers existing defects only, the pigs must be run at sufficient intervals to detect serious defect formation before they become critical.

While promising, the technology is arguably inexact, requiring experienced personnel to obtain most meaningful results. The pigs cannot accommodate all pipeline system designs—there are significant restrictions on minimum pipe diameter, pipe shape, and radius of bends. Both pigs have difficulties in detecting certain types of problems.

When the evaluator is assured that the technique used provided meaningful results (95% detection of all defects that could have a short-term impact on line integrity, would be a reasonable expectation), he can award points based upon the timing of the pig run:

8 − (years since inspection) = point value

Example:

An instrumented pig internal inspection was performed six years ago. Test digs verified that the pig data was accurate. Only minor defects were detected and subsequently corrected. Defects as small as 5% of the wall thickness were reliably found. Defect sizes 20% of the wall thickness and larger would be considered critical. Points awarded are:

8 − 6 = 2 *points*

In the future, the weighting of this technique's impact on the risk picture will surely need to be reconsidered. The use of intelligent pigs will someday be a quite comprehensive inspection technique not only for corrosion defects, but any type of anomaly on the line. As such, it will play a significant role in risk reduction. The suggested weighting above reflects some of the present limitations of the technique.

CHAPTER FIVE

Design Index

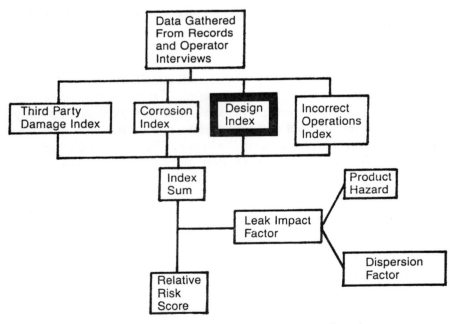

Figure 5-1. Components of the risk rating flowchart.

Design Risk

A. Pipe Safety Factor	0–25 pts	25%	(p. 107)
B. System Safety Factor	0–20 pts	20%	(p. 114)
C. Fatigue	0–15 pts	15%	(p. 117)

Design

Another significant element in the risk picture is the relation between how the pipeline was originally designed and how it is presently being operated. While this may seem straightforward, it is actually quite complex. All original designs are based upon calculations that must, for practical reasons, incorporate assumptions. These assumptions include material strengths and the use of simplifying models. Safety factors and conservativeness in the assumptions compensate for this variability caused by assumptions but cloud the view of exactly how much stress the design can safely tolerate. Further complications arise with the uncertainties in estimating existing conditions such as soil strength and actual stress loadings imposed upon the structure. In aggregate, then, the evaluator will always be uncertain in his estimation of the margin of safety.

This uncertainty should be acknowledged, but not necessarily quantified. An evaluation system should incorporate all known information and treat all unknowns (and 'unknowables') consistently. Because a relative risk picture is sought, the consistency in treatment of design variables provides a consistent base with which to perform risk comparisons.

Although this section is titled *Design*, many of the factors here are actually operating conditions. *Design* is used as an index title because all operations should be within design considerations. This section, therefore, assesses the operating environment against the critical design parameters. Many of the items evaluated here apply across section lines. The evaluator may wish to assess a pipeline system as a whole with respect to this index and the *Incorrect Operations Index.*

A. Pipe Safety Factor   **Suggested weighting 25%**
                                    Attribute

Most pipeline systems allow for some extra wall thickness in the pipe. This is normally because of the availability of standard manufactured wall thicknesses. Such "off the shelf" pipe is often more economical even though it contains more material than may be required for the intended service. This extra thickness will provide some additional protection against corrosion and external damage. This extra protection, beyond the design requirements, should be considered in the risk evaluation.

When evaluating a variety of pipe materials, distinctions in material strengths should be made. In terms of external damage protection, a tenth inch of steel offers more than does a tenth inch of fiberglass. The evaluator must make this distinction when it is desired to compare the risks associated with pipelines constructed of different materials.

This section, *Pipe Safety Factor*, is more technical than most of the other components of the evaluation. If the evaluator does not possess expertise in matters of pipeline design, outside help may be beneficial. This is not a requirement, though. By making some conservative assumptions and being consistent, a non-expert can do a credible job here. He must, however, be able to obtain some calculated values. Where original design calculations are available, few additional calculations are needed.

The procedure here is to calculate the required pipe wall thickness and compare it to the actual wall thickness (Figure 5-2). The calculated value should probably NOT include standard safety factors. This is done not only for simplicity, but also because some of the reasons for the safety factors are addressed in other sections of this risk analysis. For instance, the DOT design safety factors for gas lines are based upon nearby population density. Population density is part of the consequences section (*Leak Impact Factor*) in this evaluation system. Consequences are examined in great detail separately from design and operation considerations.

The comparison between the actual and the required wall thickness is done by using a ratio of the two numbers. Using a ratio provides a numerical scale. If this ratio is less than one, the pipe does not meet the design criteria—there is less actual wall thickness than is required by design calculations. The pipeline system has not failed either because it has not yet been exposed to the maximum design conditions, or because some error in the calculations or associated assumptions has been made. A ratio greater than one means that extra wall thickness (above design requirements) exists. For instance, a ratio of 1.1 means

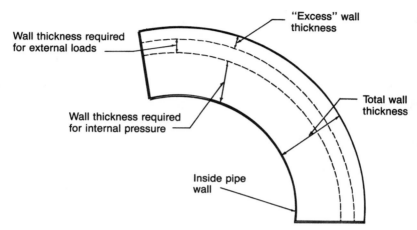

Figure 5-2. Cross section of pipe wall illustrating the pipe safety factor.

that there is 10% more pipe wall material than required by design and 1.25 means 25% more material.

Calculation of the required wall thickness involves several steps. First, Barlow's formula for circumferential stress is used to determine the minimum wall thickness required for internal pressure alone. This calculation is demonstrated in Appendix C. Barlow's calculation assumes a uniform material thickness and requires the input of a maximum allowable stress. It yields a stress value for the extreme fibers of the pipe wall (for the stress due solely to internal pressure). By starting with a maximum allowable material stress, the wall thickness needed to contain a given pressure is calculated. Alternately, inputting a wall thickness into the equation yields a maximum internal pressure that the pipe can withstand.

Depending on the method of manufacture, the assumption of uniform material may not be valid. If this is the case, the maximum allowable stress value must reflect the true strength of the material. In the case of longitudinally welded steel pipe, for instance, the weld seam and area around it are metallurgically different from the parent steel. If it is thought that such seams weaken the pipe wall, the minimum calculated wall thickness must be increased to allow for the weakness. Evidence exists that suggests that electrical resistance welded (ERW) pipe manufactured before 1970 is more prone to failure than ERW pipe manufactured after 1970 or seamless pipe or pipe manufactured by other methods. A derating factor for pre-1970 ERW pipe might be warranted.

If data from recent pressure tests verifies the material allowable stress, such tests can be considered to be evidence, but should not be considered to be conclusive proof. A history of failures that are attributable in part or in whole to a specific pipe manufacture process is sufficient reason to question the allowable stress level of the pipe, regardless of pressure test results. Favorable pressure test results will still affect the risk picture in the Hydrostatic Test item later.

In the absence of reliable, recent pressure test data and especially if the material ratings are questioned, the maximum pressure to which the pipe has been subjected (usually the pre-service hydrostatic test) can be used to calculate a material allowable stress. That is, input the maximum internal pressure into Barlow's formula to calculate a material allowable stress value. From this allowable stress value, a minimum required wall thickness can then be calculated.

Once the required wall thickness for internal pressure has been established, other loadings to which the pipe will be subjected must be considered. These other loadings include the weight of the soil over the buried line, the loadings caused by traffic moving over the line, possible soil movements (settling, faults, etc.), water pressures for submerged lines, and pipe weight. If detailed calculations are not deemed to be cost effective, the evaluator may use a standard percentage to add to the wall thickness required for internal pressure to account for all other loadings combined. Ten percent or twenty percent additional wall thickness would be conservative for steel pipe under normal loading conditions, for instance. This percentage should be increased for sections that may be subjected to additional loadings. For instance, uncased pipe under roadways would require additional wall thickness to handle the increased loads. Rigid pipe also requires more wall thickness to support external loads than does flexible pipe.

Often, casing pipe is installed to carry anticipated external loads. A casing pipe is merely a pipe larger in diameter than the carrier pipe whose purpose is to protect the carrier pipe from external loads (Figure 4-2). Casing pipe has been shown to cause cathodic protection problems for pipelines. The effect of casings on the risk picture from a corrosion standpoint is covered in the *Corrosion Index*. The impact on the *Design Index* is found here, when the casing carries the external load and allows a higher pipe safety factor for the section being evaluated. (See The Case For/Against Casings, p. 69.)

The anticipated conditions under which the line will operate should also be factored in here. The maximum allowable stress is dependent upon the temperature. Hence temperature extremes may require different wall thicknesses. Cyclic loadings and fatigue should be a consideration in material selection and wall thickness determination. Surge (water hammer) pressures should also be included in maximum pressure determination. See Appendix C for a brief review of wall thickness design considerations.

In composite pipelines, many more complexities are introduced. Often used to handle more corrosive materials, such composites may have a layer of corrosion or chemical degradation resistant material and a layer of higher strength (structural) material. Because two or more materials are involved, the stresses in each and the interaction effects must be understood. Such calculations are not easily done. Original design calculations must be used (or re-created, when not available) to determine minimum required wall thicknesses. The evaluator must then be sure that additional wall thickness of one or more of the materials will indeed add to the pipe strength and corrosion resistance, and not detract from it. It is conceivable that an increase in wall thickness in one layer may have an undesirable effect on the overall pipe structure. Further, some materials may allow diffusion of the product. When this occurs, composite designs may be exposed to additional stresses.

When all of these factors have been considered, a simple point schedule can be employed to award points based upon how much extra wall thickness exists. This schedule uses the ratio of actual pipe wall to pipe wall required and calls this ratio, t. Note that the actual pipe wall thickness is NOT the nominal wall thickness. Nominal wall thicknesses are used by manufacturers to designate a wall thickness plus or minus a manufacturing tolerance. For the purposes of this assessment, the lowest actual wall thickness in the section must be used. If actual thickness measurement data is not available, the nominal wall thickness *minus* the maximum manufacturing tolerance can be used.

t	Points
<1.0	−5 WARNING
1.0–1.1	2
1.11–1.20	5
1.21–1.40	9
1.41–1.60	12
1.61–1.80	16
>1.81	20

A simple equation can also be used instead of this table:

$(t - 1) \times 20 = $ point value

yields approximately the same values and has the benefit of more discrimination between differences in t.

Some examples to illustrate the pipe safety factor follow:

Example A:

A cross-country steel pipeline is being evaluated. The line transports natural gas. Original design calculations are available. The evaluator feels that no extraordinary conditions exist on the line and proceeds as follows:

1. He uses information from the design file to determine the required wall thickness. A MAOP of 2,000 psig using a grade of steel rated for 35,000 psi maximum allowable stress yields a wall thickness of 0.60 inches for this diameter of pipe (see Appendix C). External load calculations show the need for an additional 0.08 inches in thickness to handle the additional stresses anticipated. Surge pressures, extreme temperatures, or other loadings are extremely unlikely. The total required wall thickness is therefore 0.60 + 0.08 = 0.68 inches.

2. The actual pipe wall thickness installed is a nominal 0.88 inches. Manufacturing tolerances allow this nominal to actually be as thin as 0.79 inches. No documented thickness readings indicate that the line is any thinner than this .079 inch value, so the evaluator uses 0.79 as the actual wall thickness.

3. The ratio of actual to required wall thickness is therefore 0.79 ÷ 0.68=1.16. There exists 0.16 inches (or 16%) of additional protection against external damage or corrosion.

4. The point value for 16% extra wall thickness is 3.2, using the equation.

Example B:

Another cross-country steel pipeline is being evaluated. Hydrocarbon liquids are being transported here. In this case, original design calculations are not available. The line is 35 years old and is exposed to varying external loadings. The evaluator proceeds as follows:

1. Because of the age of the line and the absence of original documents, the most recent hydrostatic test pressure is used to determine the maximum allowable stress for the pipe material. Using the test pressure of 2,200 psig, the stress level is calculated to be 27,000 psi (see Appendix C). The evaluator is thus reasonably sure that the pipeline can withstand a stress level of 27,000 psi. The maximum allowable operating pressure of the line is 1,400 psig. Using this value and a stress level of 27,000 psi, the required wall thickness (for internal pressure only) is calculated to be 0.38 inches.

2. Using some general calculations and the opinions of the design department, the evaluator feels that an additional 10% must be added to the wall thickness to allow for external loadings for most conditions. This is an additional .04 inches. He adds an additional 5% (total of 15% above requirements for internal pressure alone) for situations where the line crosses beneath roadways. This 5% is thought to account for all types of uncased road crossings, regardless of pipeline depth, soil type, roadway design, and traffic speed and type. In other words, 15% wall thickness above that required for internal pressure only is the requirement for the worst case situation. This is an additional 0.06 inches for sections that have uncased road crossings.

3. Water hammer effects can produce surge pressures up to 100 psig. Such surges could lead to an internal pressure as high as 1,500 psig (100 psig above MAOP). This additional pressure requires an additional 0.02 inches of wall thickness.

4. The required minimum wall thicknesses are therefore 0.38 + 0.06 + 0.02 = 0.46 inches for sections with uncased crossings, and 0.38 + 0.04 + 0.02 = 0.44 inches for all other sections.

5. The evaluator next determines the actual wall thickness. Records indicate that the original purchased pipe had a nominal wall thickness of 0.55 inches. When the manufacturing tolerance is subtracted from this, the wall thickness is 0.51 inches. Field personnel, however, mention that wall thickness checks have revealed thicknesses as low as 0.48 inches. This is confirmed by documents in the files. The evaluator chooses to use 0.48 inches as the actual wall thickness, because this is the worst case expected.

6. The actual-to-required wall thickness ratios are therefore 0.48 / 0.46 = 1.04 and 0.48 / 0.46 = 1.09 for sections with and without uncased road crossings respectively. These ratios yield point values of 0.8 and 1.8 respectively. Conservatism requires that the evaluator assign a value of 0.8 points for this section of pipeline.

B. System Safety Factor   **Suggested weighting......... 20%**
                                          Attribute

Another general consideration in this section is the difference between the design pressure and the present operating pressure. In *Pipe Safety Factor* this was analyzed in terms of the pipe wall thickness alone. Here, all components of the pipeline system are included. It is a quick measure of how the system CAN be operated versus how it is presently being operated. A system being operated at its limit, leaves no room for error. Where a margin or safety factor exists, risk is reduced.

System Safety Factor fits the definition of an attribute. It is not easy to change because neither the system MAOP nor the design pressure are normally variable. Where changing either of these is an option, risk can be significantly reduced.

Each pipeline component has a specified maximum operating pressure. This value is given by the manufacturer or determined by calculations. The lowest pressure rating in the system determines the weakest component and is used to set the design pressure. Ideally, the design pressure as it is used here, should not include safety factors for the individual components. It may be difficult, however, to separate the safety factor from the actual pressure-containing capabilities of the component.

A flange, for instance, may be rated by the manufacturer to operate at a pressure of 1,400 psig. It can be tested for short periods at pressures up to 2,160 psig. It is not obvious exactly how much pressure the flange can withstand from these numbers and it is a non-trivial matter to calculate it. For purposes of this risk assessment, the value of 1,400 psig should probably be used as the maximum flange pressure even though this value certainly has a safety factor built in. The separation of the safety factor would most likely not be worth the effort.

A pressure vessel, on the other hand, normally has its design calculations available. This would allow easy separation of the safety factor. Again, if these calculations are not available, the best course is probably to use the rated operating pressure. This will yield the most conservative answer. Again, consistency is important.

As in the Pipe Safety Factor analysis, a ratio is used to show the difference between what the system CAN do and what it is presently being asked to do. This ratio will be called the *design-to-MAOP* ratio and should be the system Maximum Allowable Operating Pressure divided by the pressure rating of the weakest component. When this ratio is equal to 1, there is no safety factor present (discounting some component safety factors that were not separated). This means that the system is being operated at its limit. If the ratio is less than 1, the system can theoretically fail at any time because there is a component of the system that is not rated to operate at the system MAOP. A ratio greater than 1, means that there is a safety factor present; the system is being operated below its limit.

A simple schedule can now be developed to assign points. It may look something like this:

Design-to-MAOP Ratio	Points
2.0	20
1.75–1.99	16
1.50–1.74	12
1.25–1.49	8
1.10–1.24	5

1.00–1.10 0
< 1.00 .... −10

An equation can also be used instead of the point schedule:

[(design-to-MAOP ratio) − 1] × 20 = points

The steps for the evaluator are therefore:

1. Determine the pressure rating of the weakest system component.
2. Divide this pressure rating (from 1.) into the MAOP.
3. Assign points based upon the schedule.

Note: No credit is given for weaker components that are protected from overpressure by other means. These scenarios are examined in detail in the *Incorrect Operations Index*. The reasoning here is that the entire risk picture is being examined in small pieces. The fact that there exists a weak component contributes to this piece of the risk picture, regardless of protective actions taken. Even though a pressure vessel is protected by a relief valve, or a thin-walled pipe section is protected by an automatic valve, the presence of such weak components in the section being evaluated causes the lower 'Design-to-MAOP' ratio and hence the lower point values. Of course, the evaluator may insert a section break if he feels that a higher pressure section is being penalized by a lower rated section when there is adequate isolation between the two. Regardless of his choice, the adequacy of the isolation will be evaluated in the *Incorrect Operations Index*.

Example A:

The evaluator is examining a section of a jet fuel pipeline. The MAOP of the pipeline is 1,200 psig. In this particular section there is an aboveground storage tank that is rated for 1,000 psig maximum. The tank is the weakest component in this section. It is located on the low pressure end of the pipeline and is protected by relief systems and redundant control valves to never see more pressure than 950 psig. This effectively isolates the tank from the pipeline system and does not require that the pipeline be downrated to a lower operating pressure. These safety measures, however, are not considered for this item and the Design-to-

MAOP ratio is: (weakest component ÷ system MAOP) = 1000/1200 = 0.80. This is based on the fact that the weakest component can withstand only 1000 psig. This rates a point score of −10 points.

Example B:

In this section, the only components are pipe and valves. The pipe is designed to operate at 2,300 psig by appropriate design calculations. The overall system is rated for a MAOP of 800 psig. The valve bodies are nominally rated for maximum pressures of 1,400 psig, with permissible hydrostatic test pressures of 2,200 psig. The evaluator rates the weakest component, the valve bodies, to be 1,400 psig. Because he has no exact information as to the strength of the valve bodies, he uses the pressure rating that is guaranteed by the manufacturer for long-term service. The design-to-MAOP ratio is therefore:

1400/800 = 1.75 which yields a point value of 15 points.

Example C:

Here, a section has valves, meters, and pipe. The MAOP is 900 psig. The pipe strength is calculated to be 1,700 psig. The valve bodies and meters can all withstand pressure tests of 2,700 psig and are rated for 1,800 psig in normal operation. Again, the evaluator has no knowledge of the exact strength of the valves and meters, so he uses the normal operation rating of 1,800 psig. The weakest component, the pipe, governs, therefore:

1700/900 = 1.89 which yields a point value of 17.8 points.

C. Fatigue   **Suggested weighting .... 15% Prevention**

Fatigue failure is the largest single cause of metallic material failure (see Keyser [16]). Because a fatigue failure is a brittle failure, it can occur with no warning and with disastrous consequences.

This item may be either a prevention or an attribute, depending upon the specific system. If it is relatively easy to change the cause of the cycling, it should be considered to be a prevention. If the cycling is a necessary part of the system operation, it is more of an attribute. Fatigue is labelled a prevention here because it is believed that, in many cases, the fatigue contributors are readily changed.

Fatigue is the weakening of a material due to repeated cycles of stress. The amount of weakening is dependent upon the number and the magnitude of the cycles. Higher stresses, occurring more often, can cause more damage to the material. Factors such as surface conditions, geometry, material processes, fracture toughness, and welding processes influence susceptibility to fatigue failure (see *Cracking: A Deeper Look* , pg. 120).

Predicting the failure of a material when fatigue loadings are involved is an inexact science. Theory holds that all materials have flaws—cracks, laminations, other imperfections—if only at a microscopic level. Such flaws are generally too small to cause a structural failure, even under the higher stresses of a pressure test. These flaws can grow though; enlarging in length and depth as loads (and hence stress) are applied and then released. After repeated episodes of stress increase and reduction (sometimes hundreds of thousands of these episodes are required), the flaw can grow to a size large enough to fail at normal operating pressures. Unfortunately predicting flaw growth accurately is not presently possible from a practical standpoint. Some cracks may grow at a controlled, rather slow rate, while others may grow literally at the speed of sound through the material. The mechanisms involved are not completely understood.

For the purposes of risk analysis, the evaluator need not be able to predict fatigue failures. He must only be able to identify, in a relative way, pipeline structures that are more susceptible to such failures.

Because it is conservative to assume that any amount of cycling is potentially damaging, a schedule can be set up to compare numbers and magnitudes of cycles. Stress magnitudes should be based on a percentage of the normal operating pressures. A one hundred psi pressure cycle will have a potentially greater effect on a system rated for 150 psi MAOP than on one rated for 1,500 psi. Most research points to the requirement of large numbers of cycles at all but the highest stress levels, before serious fatigue damage occurs.

In many pipeline instances, the cycles will be due to changes in internal pressure. The following example schedule is therefore based on internal pressures as percentages of MAOP. If another type of loading is more severe, a similar schedule can be developed. Stresses caused by vehicle traffic over a buried pipeline would be an example of a cyclic loading that may be more severe than the internal pressure cycles.

This is admittedly an oversimplification of this complex issue. Fatigue is dependent upon many variables including temperature, type of stress, surface condition, and geometry of the structure. At certain stress levels, even the frequency of cycles—how fast they are occurring—are found to affect the failure point. For purposes of this assessment, however, the fatigue failure risk is being reduced to the two variables of stress magnitude and number. The following schedule is offered as a possible simple way to evaluate fatigue's contribution to the risk picture.

Table 5-1

	Lifetime Cycles				
%MAOP	$<10^3$	10^3-10^4	10^4-10^5	10^5-10^6	$>10^6$
100	7	5	3	1	0
90	9	6	4	2	1
75	10	7	5	3	2
50	11	8	6	4	3
25	12	9	7	5	4
10	13	10	8	6	5
5	14	11	9	7	6

One cycle is defined as going from the starting pressure to a peak pressure and back down to the starting pressure. The peak is measured as a percentage of MAOP.

The evaluator uses this table to analyze various combinations of pressure magnitudes and cycles. The point value is obtained by finding the worst case combination of pressures and cycles. This worst case is the situation with the lowest point value. Note the "equivalents" in this table; 9 thousand cycles at 90% of MAOP is thought to be the equivalent of 9 million cycles at 5% of MAOP; 5,000 cycles of 50% MAOP is equal to 50,000 cycles at 10% of MAOP, etc. In moving around in this table, the upper right corner is the condition with the greatest risk, and the lower left is the least risky condition. The upper left corner and the lower right corner are roughly equal.

Note also that the table is not linear. The designer of the table did not change point values proportionally with changes in either the magnitude or frequency of cycles. This indicates a belief that changes within certain ranges have a greater impact on the risk picture.

The following example illustrates further the use of this table.

Example:

The evaluator has identified two types of cyclic loadings in a specific pipeline section: 1) a pressure spike of about 200 psig caused by the start of a compressor about twice a week, and 2) vehicle traffic causing a 5 psi external stress at a frequency of about 100 vehicles per day. The section is approximately 4 years old and has an MAOP of 1,000 psig. The traffic loadings and the compressor loadings have both been occurring since the line was installed.

For the first case, the evaluator enters the table at (2 starts/week × 52 weeks/year × 4 years) = 416 cycles across the horizontal axis, and (200 psig/1000 psig) = 20% of MAOP on the vertical axis. This combination yields a point score of about 13 points.

For the second case, the lifetime cycles are equal to (100 vehicles/day × 365 days/year × 4 years) = 146,000. The magnitude is equal to (5 psig/1000 psig) = 5%. Using these two values, the schedule assigns a point score of 7 points.

The worst case, 7 points, is assigned to the section.

Cracking: A Deeper Look . . .

As contributors to fatigue failures, several common initiating mechanisms have been identified. Hydrogen-induced cracking (HIC), stress corrosion cracking (SCC), and sulfide stress corrosion cracking (SSCC) are recognized flaw creating or propagating phenomena (see *Corrosion Index*). The susceptibility of a material to these mechanisms is dependent upon several variables. The material composition is one of the more important variables. Alloys, added in small quantities to iron-carbon mixtures, create steels with differing properties. Toughness is the material property that resists fatigue failure. A trade-off often occurs as material toughness is increased, but other important properties such as corrosion-resistance, weldability, brittle-ductile transitions may be adversely affected. The fracture toughness of a material is a measure of the degree of plastic deformation which can occur before full failure. This plays a significant role in fatigue failures. Much more energy is required to fail a material that has a lot of fracture toughness, because the material can absorb some of the energy that may otherwise be contributing directly to a failure. A larger defect is required to fail a material having fracture toughness. Compare glass (low fracture toughness) with copper (high fracture toughness). In general, as yield strength goes up, fracture toughness goes down. Therefore, flaw tolerance often decreases in higher strength materials.

Another contributor to fatigue failures is the presence of stress concentrators. Any geometric discontinuity such as a hole, a crack, or a notch, can amplify the stress level in the material. Coupled with the presence of fatigue loadings, the situation can be further aggravated and make the material even more susceptible to this type of failure.

The process of heating and cooling of steel during initial formation and also during subsequent heating (welding), plays a large role in determining the microstructure of the steel. The microstructure of two identical compositions that were heat treated in different manners, may be completely different. One may be brittle (lacks toughness), and the other might be ductile at normal temperatures. The welding process forms what is known as the heat affected zone (HAZ). This is the portion of the parent metal adjacent to the weld that has an altered microstructure due to the heat of the welding operation. The HAZ is often a more brittle area in which a crack initiates.

Because the HAZ is an important element in the structural strength of the pipe, special attention must be paid to the welding process that creates this HAZ. The choice of welding temperature, speed of welding, pre-heating, post-heating, weld metal type, and even the type of weld flux, all affect the creation of the HAZ. Improper welding procedures, either because of the design or execution of the welding, can create a pipeline that is much more susceptible to failure due to cracking. This element of the risk picture is considered in the potential for human error in *Incorrect Operations Index*.

So-called avalanche fractures, where crack propagation extends literally for miles along the pipeline, have been seen in large diameter, high pressure gas lines. In these scenarios, the speed of the crack growth exceeds the pipeline depressurization wave. This can lead to a violent pipe failure where the steel is literally flattened out or radically distorted for great distances. From a risk standpoint, such a rupture extends the release point along the pipeline, but probably does not materially affect the amount of product released. An increased threat of damage due to flying debris is present. Preventive actions to this type of failure include crack arresters—sleeves or other attachments to the pipe designed to slow the crack propagation until the depressurization wave can pass—and the use of more crack resistant materials including multilayer wall pipe. If the evaluator is particularly concerned with this type of failure and feels that it can increase the risk picture in his systems, he can adjust the hazard score in the fatigue analysis by giving credit for crack arrester installations, and recognizing the susceptibility of large diameter, high pressure gas lines (particularly those lacking material toughness) by reducing point scores.

It has been proposed that fracture mechanics, allowing for all these variables, would be a better criteria than the present rejection criteria for pipeline flaws. Present regulations require the conservative rejection of welds and steels with cracks of certain lengths, regardless of depth. This is partly due to the use of radiography for inspection of steel. Radiography is sensitive to crack length but not depth. An inspection technique such as ultrasound would be required to determine all flaw dimensions. This determination would be required before a fracture mechanics rejection criteria could be used. Such a switch in inspection criteria would likely impact the risk picture.

D. Surge Potential   Suggested weighting......... 10%
                                    Prevention

The potential for pressure surges, or "water hammer" effects, is assessed here. The common mechanism for surges is the sudden conversion of kinetic energy to potential energy. A mass of flowing fluid in a pipeline, for instance, has a certain amount of kinetic energy associated with it. If this mass of fluid is suddenly brought to a halt, the kinetic energy is converted to potential energy in the form of pressure. A sudden valve closure is a common initiator of such a pressure surge or, as it is sometimes called, a pressure spike. A moving product stream contacting a stationary mass of fluid (while starting and stopping pumps, perhaps) is another possible initiator.

This pressure spike is not isolated to the region of the initiator. It forms a pressure wave that travels back upstream along the pipeline, ADDING to the static pressure already in the pipeline. A pipeline with a high upstream pressure might be overstressed as this pressure wave arrives, causing the total pressure to exceed the MAOP.

The magnitude of the pressure surge is dependent upon the fluid modulus (density and elasticity), the fluid velocity, and the speed of flow stoppage. In the case of a valve closure as the flow stoppage event, the critical aspect of the speed of closure might not be the total time it takes to close the valve. Most of the pressure spike occurs from the last 10% of the closing of a gate valve, for instance.

From a risk standpoint, the situation can be improved through the use of surge protection devices or devices that prevent quick flow stoppages (such as valves being closed too quickly). The operator must understand the hazard and all possible initiating actions, before corrective measures can be correctly employed. The evaluator should assure himself that the operator does indeed understand surge potential (see Appendix D for calculations formulae). He can then assign points to the section based upon the chances of a hazardous surge occurring.

To simplify this process, it is recommended that a hazardous surge be defined as one which is greater than 10% of the pipeline MAOP. It may be argued in some cases that a line, in its present service, may operate far below MAOP and, hence, a 10% surge will still not endanger the line. A valid argument, perhaps, but perhaps also an unnecessary complication in the risk assessment. The evaluator should decide on a method and then apply it uniformly to all sections being evaluated.

The point schedule can be set up with three general categories and room for interpolation between the categories:

Evaluate the chances of a pressure surge of magnitude greater than 10% of system MAOP:

High probability .... 0 pts
Low probability .... 5 pts
Impossible .... 10 pts

High probability exists where closure devices, equipment, fluid modulus, and fluid velocity all support the possibility of a pressure surge. No mechanical preventors are in place. Operating procedures to prevent surges may or may not be in place.

Low probability exists when surges can happen (fluid modulus and velocity can produce the surge), but are safely dealt with by mechanical devices such as surge tanks, relief valves, slow valve closures, etc., in addition to operating protocol. Low probability also exists when the chance for a surge to occur is only through a rather unlikely chain of events.

Impossible means that the fluid properties cannot, under any reasonable circumstances, produce a pressure surge of magnitude greater than 10% MAOP.

Example:

A crude oil pipeline has flow rates and product characteristics that are supportive of pressure surges in excess of 10% of MAOP. The only identified initiation scenario is the rapid closure of a mainline gate valve. All of these valves are equipped with automatic electric openers that are geared to operate at a rate less than the critical closure time (see Appendix D). If a valve must be closed manually, it is still not possible to close the valve too quickly—many turns of the valve handwheel are required for each 5% valve closure. Points for this scenario are assessed at 5.

E. System Hydrostatic Test   Suggested weighting......... 25%
                                   Prevention

A hydrostatic test is a pressure test in which the pipeline is filled with water, then pressurized to a predetermined pressure, and held at this test pressure for a predetermined length of time. This test pressure normally exceeds the anticipated maximum internal pressure. It is a powerful technique in that it proves the strength of the entire system. The hydrostatic test is perhaps the ultimate inspection tool. It provides virtually indisputable evidence as to the system integrity (within the test parameters).

All materials have flaws and defects, if only at the microscopic level. Given enough stress, any crack will enlarge, growing in depth and width. Crack growth is not predictable. It may occur gradually or literally at the speed of sound through the material. Under the constant stress of a hydrostatic test, it is reasonable to assume that a group of flaws beyond some minimum size are growing. Below this minimum size, cracks will not grow unless the stress level is increased. If the stress level is rather low, only the largest of cracks will be growing. At higher stresses, smaller and smaller cracks will begin to grow, propagating through the material. When a crack reaches a critical size at a given stress level, rapid, brittle failure of the structure is likely. (See explanations of fracture toughness and crack propagation in *Corrosion Index* and *Design Index*).

By conducting a hydrostatic test at high pressures, the pipeline is being subjected to stress levels higher than it should ever encounter in everyday operation. Ideally, then, when the pipeline is de-pressured from the hydrostatic test, the only cracks left in the material are of a size that will not grow under the stresses of normal operations. All cracks that could have grown to a critical size under normal pressure levels would have already grown and failed under the higher stress levels of the hydrostatic test.

Research suggests that the length of time that a test pressure is maintained is not a critical factor. This is based upon the assumption that there is always crack growth and whenever the test is stopped, a crack might be on the verge of its critical size, and hence, failure.

The pressure level, however, is an important parameter. The higher the test pressure relative to the normal operating pressure, the greater the safety margin. The chances of a pressure reversal, in which a pipeline fails at a pressure less than the test pressure, becomes increasingly remote as the margin between test and operating pressures increases. This is explained by the theory of critical crack size discussed above. A hydrostatic test does not last forever. Corrosion, third party damages, soil movements, pressure cycles, etc., all contribute to the constantly changing risk picture. A pipeline should be retested at appropriate intervals to prove its structural integrity.

Interpretation of hydrostatic test results is a non-trivial exercise. Although time duration of the test may not be critical, the pressure is normally maintained for at least four hours for practical reasons, if not for compliance with applicable regulations. During the test time (which is often 4 to 24 hours), temperature and strain will be affecting the pressure reading. It requires a knowledgeable test engineer to properly interpret pressure fluctuations and to distinguish between a transient effect and a small leak on the system or the inelastic expansion of a component.

The point schedule for hydrostatic testing can assume proper test methods, and assess the impact on risk on the basis of time since the last test and the test level (in relation to the normal maximum operating pressures). An example schedule follows:

a) Calculate H, where H = (Test Pressure/MAOP)

> H < 1.10 (1.10= test pressure 10% above MAOP) ... 0 pts
> 1.11 < H < 1.25 .... 5 pts
> 1.26 < H < 1.40 .... 10 pts
> H > 1.41 .... 15 pts

or a simple equation can be used:

> (H − 1) × 30 = point score
> min = 0 points

b) Time since last test: Points = 10 − (years since test)

> A test four years ago .... 6 pts
> A test eleven years ago .... 0 pts
> min = 0 points

Add points from (a) and (b) above for total hydrostatic test score.

In this schedule, maximum points are given to a test which occurred within the last year and which was to a pressure greater than 40% above the maximum operating pressure.

Example:

The evaluator is studying a natural gas line whose MAOP is 1,000 psig. This section of line was hydrostatically tested six years ago to a pressure of 1,400 psig. Documentation on hand indicates that the test was properly performed and analyzed. Points are awarded as follows:

H = 1400/1000 =1.4

a) (1.4 − 1) × 30 .... 12 pts
b) 10 − 6 years .... 4 pts
12 + 4 .... 16 pts

F. Soil Movements   **Suggested weighting 5%**
                         Attribute

Under certain conditions, the pipeline may be subjected to stresses due to soil movements. These movements may be sudden and catastrophic or they may be long-term deformations that induce stresses on the pipeline over a period of years. These can add considerable stresses to the pipeline and should be carefully considered in a risk analysis. While soil movements are included as a component of the pipe wall thickness determination, the pipe itself cannot always be designed to withstand the movements. Therefore, in this item, the potential for these pipe stresses along with remedial measures is assessed.

Many, if not most, of the potentially dangerous soil movement scenarios have a slope involved (Figure 5-3). The presence of a slope adds the factor of gravity. Landslides, mudflows, and creep are the more well known downslope movements phenomena. Another movement involving freezing, thawing, and gravity is solifluction, a cold-regions phenomenon distinct from the more common movements [32].

Effects that are not slope-oriented include soil swelling and shrinkage. These can be caused by differential heating, cooling, or moisture contents. Sudden subsidence can cause shear forces as well as bending stresses.

Frost heave is another cold-region phenomenon involving temperature and moisture effects that cause soil movements. As ice or ice lenses are formed in the soil, the soil expands due to the freezing of the moisture. This expansion can cause vertical or uplift pressure on a buried pipeline. The amount of increased load on the pipe is partially dependent upon the depth of frost penetration and the pipe characteristics. Rigid pipes are more easily damaged by this phenomenon.

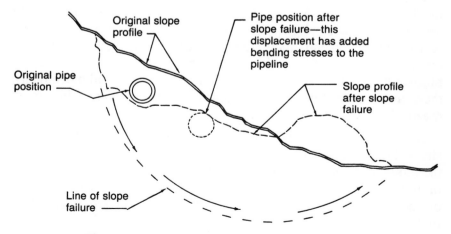

Figure 5-3. Sudden slope failure over pipeline.

Pipelines are generally placed at depths below the frost lines to avoid frost loading problems.

Many pipelines traverse areas of highly expansive clays that are particularly susceptible to swelling and shrinkage due to moisture content changes. These effects can be especially pronounced if the soil is confined between non-yielding surfaces. Such movements of soil against the pipe can damage the pipe coating as well as induce stresses in the pipe wall. Good installation practice avoids embedding pipes directly in such soils. A bedding material is used to surround the line to protect the coating and the pipe. Again, rigid pipes are more susceptible to structural damage from expansive soils.

A geo-technical evaluation is the best method to determine the potential for significant ground movements. In the absence of this however, the evaluator should seek out evidence in the form of operator experience. Large cracks in the ground during dry spells, sink holes or sloughs in periods of heavy rain, foundation problems on buildings nearby, landslide or earthquake potential, observation of soil movements over time or on a seasonal cycle, and displacements of buried structures discovered during routine inspections are all indicators that the area is susceptible. Even a brief survey of the topography together with information as to the soil type and the climatic conditions should either readily confirm the operator's experience or establish a doubt in the evaluator's mind.

Anticipated soil movements are often confirmed by actual measurements. Instruments such as inclinometers and extensometers can be used to detect even slight soil movements. While these instruments reveal soil movements, they are not necessarily a direct indication of the stresses induced on the pipe. They only indicate increased probability of additional pipe stress. In areas prone to soil movements, these instruments can be set to transmit alarms to warn when more drastic changes have occurred.

Movements of the pipe itself are the best indication of increased stress. Strain gauges attached to the pipe wall can be used to monitor the movements of the pipeline, but must be placed to detect the areas of greatest pipe strain (largest deflections). This requires a knowledge of the most sensitive areas of the pipe wall and the most likely movement scenarios. Use of these gauges provides a direct measure of pipeline strain that can be used to calculate increased stress levels.

The evaluator can establish a point schedule for assessing the risk of pipeline failure due to soil movements. The point scale should reflect the relative risk among the pipeline sections evaluated. If the evaluations cover everything from pipelines in the mountains of Colorado to the deserts of the Middle East, the range of possible point values should similarly cover all possibilities. Evaluations performed on pipelines in a consistent environment may need to incorporate more subtleties to distinguish the differences in risk.

The following schedule is designed to cover pipeline evaluations in which the pipelines are in moderately differing environments.

Potential for significant (damaging) soil movements:

High .... 0 pts
Medium .... 2 pts
Low .... 6 pts
None ....10 pts
Unknown .... 0 pts

High. Areas where damaging soil movements are common or can be quite severe. Regular fault movements, landslides, subsidence, creep, or frost heave are seen. The pipeline is exposed to these movements. A rigid pipeline in an area of less frequent soil movements should also be classified here due to the increased susceptibility of rigid pipe to soil movement damage. Active earthquake faults in the immediate vicinity of the pipeline should be included in this category.

Medium. Damaging soil movements are possible but rare or unlikely to affect the pipeline due to its depth or position. Topography and soil types are compatible with soil movements, although no damage in this area has been recorded.

Low. Evidence of soil movements is rarely if ever seen. Movements and damage are not likely. There are no recorded episodes of structural damage due to soil movements. All rigid pipelines should fall into this category as a minimum, even when movements are rare.

None. No evidence of any kind to indicate potential threat due to soil movements.

Corrective actions can be performed to the point at which the potential for significant movements is none. Examples include dewatering of the soil using surface and subsurface drainage systems, and permanently moving the pipeline. While changing the moisture content of the soil does indeed change the soil movement picture, the evaluator should assure himself that the potential has in fact been eliminated and not merely reduced, before he assigns the "none" classification. Moving the line includes burial at a depth below the movement depth (determined by geotechnical study; usually applies to slope movements), moving the line out of the area where the potential exists, and placing the line above-ground (may not be effective if the pipe supports are subject to soil movement damage).

Where a potential exists, point values may be adjusted by the following mitigative actions:

Monitoring at least annually .... +1 pts
Continuous monitoring .... +2 pts
Stress relieving .... +3 pts
Adjust points to a maximum of 10 points.

Continuous monitoring offers the benefit of immediate indication of potential problems. This can be accomplished by transmitting a signal from a soil movement indicator or from strain gauges placed on the pipeline. Proper response to these signals is implied in awarding the point values. Periodic surveys are also commonly used to detect movements. Surveying can not be relied upon to detect sudden movements in a timely fashion.

Stress relieving is normally accomplished by opening a trench parallel to or over the pipeline. This effectively unloads the line from any soil movement pressures that may have been applied. Another method is to excavate the pipeline and leave it above ground. Either of these is normally only a short term solution. Installing the pipeline above ground, on supports can be a permanent solution, but as already pointed out, may not be a good solution if the supports are susceptible to soil movement damage.

Example:

In the section being evaluated, a brine pipeline traverses a relatively unstable slope. There is substantial evidence of slow downslope movements along this slope although sudden, severe movements have not been observed. The line is thoroughly surveyed annually, with special attention paid to potential movements. The evaluator scores the hazard as somewhere between 'high' and 'medium' because potentially damaging movements can occur but have not yet been seen. This equates to a point score of 1 point. The annual monitoring increases the point score by one point, so the final score is 2 points.

CHAPTER SIX

Incorrect Operations Index

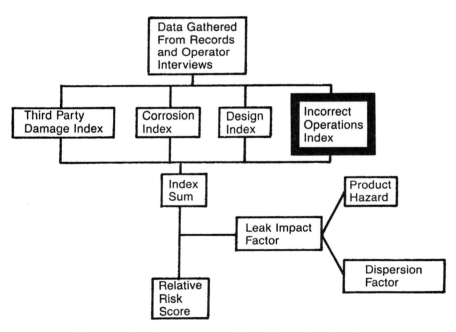

Figure 6-1. Components of the risk rating flowchart.

Incorrect Operations Risk

A. Design.......................... 30%   0–30 pts (p. 135–147)
    Hazard Identification .... 0–4 pts (p. 135)
    MAOP Potential .... 0–12 pts (p. 136)

Incorrect Operations

Human Error Potential

One of the most important aspects of risk is the potential for human error. Unfortunately, this is also perhaps the most difficult aspect to quantify, or even to understand. Safety professionals are emphasizing behavior as perhaps the key to a breakthrough in accident prevention. The factors underlying behavior and attitude cross into areas of psychology, sociology, biology, etc., and are far beyond the simple assessment technique that is being built here. It is left to the evaluators to incorporate additional knowledge and experience into this index as such knowledge becomes available. When data can statistically prove correlations between accidents and variables such as years of experience, or time of day, or level of education, or diet, or salary, then these variables can be included in the risk picture. It is not thought that the state of the art has advanced to that point yet.

In this index, the potential for human error in pipeline operations is assessed. We limit our assessment of this potential to those errors committed by the pipeline operators themselves. Vandalism or accidents caused by the public are not considered here. These are addressed to some extent in the *Third Party Damage Index.*

Human errors are estimated to have caused 62% of all hazardous materials accidents in the U.S. [28]. The public is especially sensitive to these types of risks. In the transportation industry, pipelines are comparatively insensitive to human interactions. Processes of moving products by rail or highway or marine are obviously more manpower intensive. Nonetheless, wherever human variability is involved to any extent, risk is impacted.

Human interaction in the pipeline system can be either positive—preventing or mitigating failures, or negative—exacerbating or initiating failures. Where efforts are made to improve human performance, risk reduction is achieved. Improvements are thought to occur through better designs of the pipeline system, development of better employees and/or through improved management programs. Such improvements are a component of risk management.

An important point in assessing human error risk is the supposition that small errors at any point in a process can leave the system vulnerable to failure at a later stage. With this in mind, the evaluator must assess the potential for human error in each of four phases in pipelining: design, construction, operation, and maintenance. A slight design error may not show up for years when it is suddenly the contributor to a failure. By viewing the entire process as a chain of interlinked steps, we can also identify possible intervention points, where checks or inspections or equipment can be inserted to avoid a human-error type failure.

Specific items and actions that are thought to minimize the potential for errors should be identified and incorporated into the risk assessment when they are properly employed. A point schedule is then used to weigh the relative impact of each item on the risk picture. Many of these evaluations will be subjective. The evaluator should take steps to ensure consistency. As with the *Design Index,* many of these items will probably be consistent across the pipeline sections.

The *Incorrect Operations Index* is thought to be composed of prevention items only. Even if a pipeline system has been acquired with little knowledge of original design, construction, or maintenance practices, the operator can take steps to gain evidence of the pipeline's history. The conservative approach is to assume worst case scenarios in the absence of data to the contrary. Again, consistency is the key.

A. Design

Suggested weighting 30%
Prevention

This is perhaps the most difficult category to analyze. Design and planning processes are not well defined and are often highly variable. The suggested approach is for the evaluator to ask for evidence that certain error-preventing actions occurred during the design phase. It would not be inappropriate to insist upon documentation for each item.

1. Hazard identification .... 4 points
2. MAOP potential .... 12 points
3. Safety systems .... 10 points
4. Material selection .... 2 points
5. Checks .... 2 points

1. Hazard Identification....................................0–4 points

The evaluator checks to see that efforts were made to identify all credible hazards associated with the pipeline and its operation. The hazard must be clearly understood before appropriate risk reduction measures can be employed. This would include all possible failure modes. Thoroughness is important. Have all initiating events been considered? temperature induced overpressure? fire around the facilities? safety device failure? (Haz-Op is a good method for identifying hazards. See Chapter 1.)

Ideally, the evaluator should see some documentation showing that a complete hazard identification was performed. In the absence of this, he can interview system experts to see if at least the more obvious scenarios have been addressed.

Points are awarded (maximum of four points) based upon the thoroughness of the hazard studies.

2. Potential for Reaching MAOP ....0–12 points

A simple answer to the question "what can go wrong?" is merely a measure of the possibility of exceeding the system Maximum Allowable Operating Pressure (MAOP). Obviously, a system where it is not physically possible to exceed the MAOP is inherently safer than where the possibility exists. The ease with which MAOP is reached must be assessed.

MAOP is the theoretical maximum internal pressure to which the pipeline can be subjected, less any appropriate safety factors. The safety factors allow for uncertainties. MAOP is determined from stress calculations. Internal pressure induces stresses in the wall of the pipe. The most severe of those stresses will be compared to the material stress limits. Material stress limits are theoretical values, confirmed (or at least evidenced) by testing, which predict the point at which the material will fail when subjected to higher stress.

Failure is usually defined (in a simplified way) as the point at which the material changes shape and does not return to its original form when the stress is removed. When this "inelastic" limit is reached, the material has been structurally altered from its original form. Internal pressure must never be high enough to cause a stress level that exceeds the material stress limit.

External forces also add stress to the pipe. These external stresses can be caused by the weight of the soil over a buried line, the weight of the pipe itself when it is unsupported, temperature changes, etc. In general, any external influence that tries to change the shape of the pipe will cause a stress. Some of these stresses are additive to the stresses caused by internal pressure. As such, they must be allowed for in the MAOP calculations. Hence, care must be taken to ensure that the pipeline will never be subjected to any combination of internal pressures and external forces that will cause the pipe material to be overstressed.

MAOP is the internal pressure component of the total stress situation in the pipe material. This means that after all other stress-causing forces are considered, the pipe material can still withstand an additional amount of stress equal to that which will be caused by the application of the internal pressure. Internal pressure normally accounts for the majority of total pipe wall stress. Consequently, it is normally closely controlled to ensure that it does not cause MAOP to be exceeded.

For purposes of this item, MAOP can incorporate any and all design safety factors, or it may exclude the operating safety factors that are mandated by government regulations. It should not exclude engineering safety factors that reflect the uncertainty and variability of material strengths and the simplifying assumptions of design formulae. As such they are reasonable limitations on operating pressure. Regulatory operating safety factors may go beyond this to allow for errors and omissions, deterioration of facilities, etc. Such allowances are certainly needed in pipeline operation, but will yield quite conservative results in this particular section. Regulatory safety factors may therefore be omitted from the MAOP calculations for this item. As with all points of this tool, such distinctions are ultimately left to the evaluator. Because a picture of risk relative to other pipelines is sought, any consistent definition of MAOP will work.

To define the ease of reaching MAOP (whichever definition of MAOP is used) a point schedule is designed to cover the possibilities. For example:

A. Routine ....**0 pts**

Definition: Where routine, normal operations could allow the system to reach MAOP. Over-pressure is prevented by procedure or safety device.

B. Unlikely ....**5 pts**

Definition: Where over-pressure can occur through a combination of procedural errors or omissions, and failure of safety devices (at least two levels of safety). For example: A pump can be run in a 'deadheaded' condition by the accidental closing of a valve, and two levels of safety system (a primary safety and one redundant level of safety) failing would overpressure the pipeline.

C. Extremely Unlikely ....**10 pts**

Definition: Where over-pressure is theoretically possible, but only through an extremely unlikely chain of events including errors, omissions, safety device failures at more than two levels of redundancy. For example: A large diameter gas line would over-pressure if a mainline valve was closed AND communications (SCADA) failed AND downstream vendors did not communicate problems AND local safety shutdowns failed, **AND** the situation went undetected for a matter of hours. Obviously, this is an unlikely scenario.

D. Impossible .... 12 pts

Definition: Where the pressure source cannot, under any conceivable chain of events, over-pressure the pipeline.

Ideally, the evaluator uses the worst credible scenario that was created in a Haz Ops type hazard identification exercise.

In studying the point schedule for ease of reaching MAOP, the "routine" description implies that MAOP can be reached rather easily. The only preventative measure may be procedural, where the operator is relied upon to operate 100% error-free, or a simple safety device that is designed to close a valve, shut down a pressure source, or relieve pressurized product from the pipeline.

If perfect operator performance and one safety device are relied upon, the pipeline owner is accepting a high level of risk of reaching MAOP. Error-free work techniques are not realistic and industry experience shows that reliance upon a single safety shutdown device, either mechanical or electronic, allows for some periods of NO over-pressure protection. Few points should be awarded to such situations.

Remember: The evaluator is making no value judgments at this stage as to whether or not reaching MAOP poses a serious threat to life or property. Such judgments will be made when the 'consequence' factor is evaluated.

The "unlikely" description, Category B, allows for redundant levels of safety devices. These may be any combination of relief valves; rupture disks; mechanical, electrical, or pneumatic shutdown switches; or computer safeties (programmable logic controllers, supervisory control and data acquisition systems, or any kind of "logic" devices that may trigger an over-pressure prevention action). The point is that at least two independently operated devices are set to prevent over-pressure of the pipeline. This allows for the accidental failure of at least one safety device, with backup provided by another.

Operator procedures must also be in place to cause the pipeline to be operated at a pressure level below the MAOP. In this sense, any safety device can be thought of as a backup to proper operating procedures. The point value of Category B should reflect the chances, relative to the other categories, of a procedural error coincident with the failure of two or more levels of safety. Industry experience shows that this is not as unlikely an occurrence as it may first appear.

Category C, "extremely unlikely," should be used for situations that are less hazardous than Category B, but still contain a theoretical chance for exceeding MAOP. As this chance becomes increasingly remote, points awarded should come closer to a Category D score.

The "impossible" description of Category D is fairly straightforward. The pressure source is deemed to be incapable of exceeding the MAOP of the pipeline under ANY circumstances. Potential pressure sources must include pumps, compressors, wells, connecting pipelines, and the often overlooked thermal sources. A pump which, when operated in a deadheaded condition, can produce 1,000 psig pressure cannot, theoretically overpressure a line whose MAOP is 1,400 psig. In the absence of any other pressure source, this situation should receive the maximum points. The potential for thermal over-pressure must not be overlooked, however. A section of liquid-full pipe may be pressured beyond its MAOP by a heat source such as sun or fire, if the liquid has no room to expand.

Further, in examining the pressure source, the evaluator may have to obtain information from suppliers as to the maximum pressure potential of their facilities. It is sometimes difficult to obtain the maximum pressure value as it must be defined for this application, assuming failure of all safety and pressure limiting devices. In the next section, a distinction will be made between safety systems controlled by the pipeline operator and those outside his direct control.

3. Safety Systems . . . . .0–10 points

Safety devices, as a component of the risk picture, are included here in the *Incorrect Operations Index* rather than the *Design Index*. This is done because it is thought that safety systems exist as a backup for the situations in which human error causes or allows MAOP to be reached. As such, they reduce the possibility of a pipeline failure due to human error. It can be argued that most of the risk picture is at least indirectly linked to human error, but safety systems are perhaps more directly linked to incorrect operations by the pipeline operator.

The risk evaluator should carefully consider any and all safety systems in place. A safety system or device is a mechanical, electrical, pneumatic, or computer-controlled device that prevents the pipeline from being overpressured. Prevention may take the form of shutting down a pressure source, or relieving pressurized pipeline contents. Common safety devices include relief valves, rupture disks, and switches that may close valves, shut down equipment, etc., based upon sensed conditions. A level of safety is considered to be any device that unilaterally and independently causes an over-pressure prevention action to be taken. When more than one level of safety exists—each level independent of the previous device and its power source—redundancy is established (Figure 6-2). Redundancy provides backup protection in case of failure of a safety device for any reason. Two, three, and even four levels of safety are not uncommon for critical pipelines.

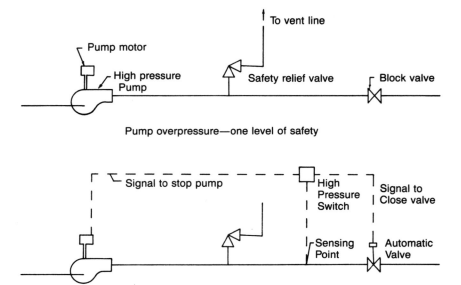

Figure 6-2. Safety systems.

In some instances, safety systems exist that are not under the direct control of the pipeline operator. When another pipeline or perhaps a producing well is the pressure source, control of that source and its associated safeties may rest with the other party. In such cases, allowances must be made for the other party's procedures and operating discipline. Uncertainty may be reduced when there is direct inspection or witnessing of the calibration and maintenance of the third party's safety equipment, but this does not replace direct control of the equipment.

A point schedule should be designed to accommodate all situations on the pipeline system*. An example schedule follows:

A. No safety devices present .... 0 pts
B. On site, one level only .... 3 pts
C. On site, two or more levels .... 6 pts
D. Remote, observation only .... 1 pts
E. Remote, observation and control .... 3 pts
F. Non-owned, active witnessing .... −2 pts
G. Non-owned, no involvement .... −3 pts
H. Safety systems not needed .... 10 pts

*Note: The evaluator must decide if he will be considering the pipeline system as a whole (ignoring section breaks) for this item. A safety system might be physically located outside of the sections it is protecting. (See Example C.).

In this example schedule, more than one safety system condition may exist at the same time. The evaluator defines the safety system and the overpressure scenarios. He then assigns points for every condition that exists. Safety systems which are not thought to adequately address the overpressure scenarios should not be included in the evaluation. Note that some conditions cause points to be subtracted.

A. No safety devices present. In this case, reaching MAOP is possible, and no safety devices are present to prevent overpressure. Inadequate, improperly designed devices would also fall into this category. A relief valve that cannot relieve enough to offset the pressure source is an example of an ineffective device. Lack of thermal overpressure protection where the need exists is another example of a situation that should receive 0 pts.

B. On site, one level. For this condition a single device, located at the site, offers protection from overpressure. The site can be the pipeline or the pressure source. A pressure switch that closes a valve is an example. A properly sized relief valve on the pipeline is another example.

C. On site, two or more levels. Here, more than one safety device is installed at the site. Each device must be independent of all others and be powered by a different power source than the others. This makes each device an independent level of safety. More points should be given for this situation because redundancy of safety devices obviously reduces risk.

D. Remote, observation only. In this case, the pressure is monitored from a remote location. Remote control is not possible and automatic overpressure protection is not present. While not a replacement for an automatic safety system, such remote observation provides some additional backup. Points may be given for such systems only if such observation is practiced 95%–100% of the time. An example would be a pressure that is monitored and alarmed in a control room which is manned 24 hours a day and which has a communication reliability rate of more than 95%. Upon notification of an abnormal condition, the observer can dispatch personnel to correct the situation.

E. Remote, observation and control. This is the same situation as the previous one with the added feature of remote control capabilities. Upon notification of overpressure, the observer is able to remotely take action to prevent overpressure. This may mean stopping a pump or compressor and opening or closing valves. Remote control capability can significantly impact the risk picture only if communications are reliable—95% or better for both receiving of the pressure signal and transmission of the control signal. Remote control generally takes the form of opening or closing valves and stopping pumps or compressors. This condition receives more points because more immediate corrective action is made possible by the addition of the remote control capabilities.

F. Non-owned, active witnessing. Here, overpressure prevention devices exist, but are not owned, maintained, or controlled by the owner of the equipment that is being protected. The pipeline owner takes steps to assure himself that the safety device(s) is properly calibrated and maintained by witnessing such activities. Inspection of calibration or inspection reports without actually witnessing the activities may, in the evaluator's judgment, earn points. Points awarded here should reflect the uncertainties attached with not having direct control of the devices. By assigning negative points here, identical safety systems under different ownerships would have different point values. This reflects a difference in the risk picture caused by the different ownerships.

G. Non-owned, no involvement. Here again, the overpressure devices are not owned, operated, or maintained by the owner of the equipment that is being protected. The equipment owner is relying upon another party for his overpressure protection. Unlike the previous category, here the pipeline owner is taking no active role in assuring himself that the safety devices are indeed kept in a state of readiness. As such, points are subtracted—the safety system effectiveness has been reduced by the added uncertainty.

H. Safety systems not needed. In the previous item, MAOP Potential, the most points were awarded for the situation in which it is impossible, under any reasonable chain of events, for the system to reach MAOP. Under this scenario, the most points are also awarded because no safety systems are needed. It is thought that this is not the normal situation in cross-country pipelines, but where it exists, this situation poses the smallest hazard.

For all safety systems, the evaluator should examine the status of the devices should a loss of power occur. Some valves and switches are designed to "fail closed" upon loss of their power supplies (electric or pneumatic, usually). Others are designed to "fail open," and a third class remains in its last position "fail last." The important thing is that the equipment fails in a mode that leaves the site in a safe condition.

Following are three examples of the application of this point schedule.

Example A:

In this pipeline section, a pump station is present. The pump is capable of overpressuring the pipeline. To prevent this, safety devices are installed. A pressure sensitive switch will stop the pump and allow product to flow around the station in a safe manner. Should the pressure switch fail to stop the pump, a relief valve will open and vent the entire pumped product stream to a flare in a safe manner. This station is remotely monitored by the transmission of appropriate data (including pressures) to a control room that is manned 24 hours per day. Remote shutdown of the pump from this control room is possible. Communications are deemed to be 98% reliable.

Conditions Present	Points
C	6
E	3
Total points = 9	

Note that two levels of safety are present (pressure switch and relief valve), and that full credit is given to the remote capabilities only after communication effectiveness is assessed.

Example B:

For this example, a section of a gas transmission pipeline has a supplier interconnect. This interconnect leads directly to a producing gas well that can produce pressures and flow rates which can overpressure the transmission pipeline. Several levels of safety are present at the well site and under the control of the producer. The producer has agreed by contract to ensure that the transmission pipeline owner is protected from any damaging pressures due to the well operation. The pipeline owner monitors flow rates from the producer as well as pressures on the pipeline. This monitoring is on a 24-hour basis, but no remote control is possible.

Conditions Present	Points
C	6
D	1
G	−3
Total Points = 4	

Note that credit is given for condition C even though the pipeline owner has no safety devices of his own in this section. The fact that the devices are present warrants points; the fact that they are not under the owner's control negates some of those points (condition G). Also, while contractual agreements may be useful in determining liabilities AFTER an accident, they are not thought to have much impact on the risk picture. If the owner takes an active role in ensuring that the safety devices are properly maintained, condition F would replace G yielding a total point score of 5.

Example C:

In this example, a supplier delivers product via a high-pressure pump into a pipeline section that relies on a downstream section's relief valve to prevent overpressure. The supplier has a pressure switch at the pump site to stop the pump in the event of high pressure. The pipeline owner inspects the pump station owner's calibration and inspection records for this pressure switch. The pump station owner remotely monitors the pump station operation 24 hours per day.

Conditions Present	Points
B	3
F-G	−2.5
Total points = 0.5	

Note that credit is not given, in this case, for a relief valve not in the section being evaluated. It could be argued that, because section breaks are somewhat arbitrarily inserted, safety device effectiveness should cross section lines. The evaluator in this case, however, has decided to strictly apply his section rule, mainly for simplicity. He saves himself many shades of gray interpretations and feels that this is worth the possible inequities that may result*.

Note also that no points are given for the supplier's remote monitoring. Again, the evaluator has made the decision to simplify—he does not wish to be evaluating suppliers' systems beyond the presence of direct overpressure shutdown devices located at the site. Finally, note that the evaluator has awarded points for the pipeline owner's inspection of the suppliers' maintenance records. He feels that, in this case, an amount of risk reduction is achieved by such inspections.

*As with every aspect of this risk assessment tool, the evaluator makes interpretations within the guidelines established. The tool can and should be customized to the particular situation. The important thing is that ALL contributors and detractors of risk are considered. The relative magnitude of each risk component has already been agreed upon by company officers and employees. Therefore, uniform application of the tool is the critical issue that the evaluator faces.

4. Material Selection .... 0–2 points

The evaluator should look for evidence that proper materials were identified and specified with due consideration to all stresses reasonably expected. This may appear to be an obvious point, but when coupled with ensuring that the proper material is actually installed in the system, a number of historical failures could have been prevented by closer consideration of this point. The evaluator should find calculations showing all anticipated stresses in the pipe components. This would include concrete coatings, internal and external coatings, nuts and bolts, all connecting systems, and supports, as well as the structural (load-bearing) members of the system. Documents should show that the corrosion potential, including incompatible material problems and welding-related problems, was considered in the design.

Most importantly, a set of control documents should exist. These control documents, in the form of pipeline specifications, give highly detailed data on all system components, from the nuts and bolts to the most complex instrumentation. The specifications will address component sizes, material compositions, paints and other protective coatings, and any special installation requirements. Design drawings specify the location and assembly parameters of each component.

When any changes are made to the pipeline, the control documents should govern. All new and replacement materials must conform to the original specifications or the specifications must be formally reviewed and revised to allow different materials. By rigidly adhering to these documents, the chance of mistakenly installing incompatible materials is reduced.

Awarding of points for this item should be based upon the existence and use of control documents that govern all aspects of pipeline material selection and installation. Two points are awarded for the best use of control documents, 0 points if control documents are not used.

5. Checks ....0–2 points

Here, the evaluator determines if design calculations and decisions were checked at key points during the design process. Ideally, a licensed professional engineer vouches for all designs. This is a definite point of intervention. Design checks by qualified professionals can help to prevent errors and omissions by the designers. Even the most routine designs require a degree of professional judgment and are consequently prone to error. Design checks can be performed at any stage of the life of the system. It is probably impossible to accurately gauge the quality of the checks—evidence that they were indeed performed will probably have to suffice.

Two points are awarded for sections whose design process was carefully monitored and checked.

B. Construction   Suggested weighting ....20 pts
                                         Prevention

Ideally, construction processes would be well-defined and invariant from site to site. In-process measurements and high pride of work-manship among the constructors would ensure the highest quality and consistency in the finished product— inspection would not be needed.

Unfortunately, this is not the present state of the art in construction practice. Conformance specifications are kept wide to allow for a myriad of conditions that may be encountered in the field. Work forces are often transient and awarding of work contracts is often done solely on the basis of lowest price. This makes the job price-driven; shortcuts are sought and speed is rewarded over attention to detail.

For the construction phase, the evaluator should find evidence that reasonable steps were taken to ensure that the pipeline section was constructed per design specifications. This includes checks on the quality of workmanship and, ideally, another check on the design phase.

While the post-construction pressure test verifies the system strength, improper construction techniques could cause problems far into the future. Residual stresses, damage to corrosion prevention systems, improper pipe support, and dents or gouges causing stress risers are some examples of construction defects that may pass an initial pressure test, but contribute to a later failure.

1. Inspection .... 10 points
2. Materials .... 2 points
3. Joining .... 2 points
4. Backfilling .... 2 points
5. Handling .... 2 points
6. Coating .... 2 points

The evaluator should satisfy himself that:

1. Inspection ....0–10 points

A qualified inspector was present to oversee all aspects of the construction. The inspection provided was of the highest quality. A check of the inspector's credentials, his notes during construction, his work history, and maybe even the constructor's opinion could be used in assessing the performance. The scoring of the following construction items may also hinge upon the inspector's perceived performance.

If inspection is a complete unknown, 0 points should be awarded. Where it is well-known and well-documented, the maximum points should be awarded. This item commands the most points for the construction item because current pipeline construction practices rely so heavily on proper inspection.

2. Materials ....0–2 points

All materials and components were verified as to their authenticity and conformance to specifications prior to construction. Awareness of potential counterfeit materials should be high for recent construction. Requisition of proper materials is not sufficient for this item. An on-site material handler should be designated to take all reasonable steps to ensure that the right material is indeed being installed in the right location.

Evidence that was properly done warrants 2 points.

3. Joining ....0–2 points

High quality of workmanship is seen in all methods of joining pipe sections. Welds were inspected by appropriate means (X-ray, ultrasound, dye penetrant, etc.) and all were brought into compliance with governing specifications. Often, weld acceptance or rejection is determined by two inspectors. Point values should be decreased for less than 100% weld inspection. Other joining methods (flanges, screwed connections, polyethylene fusion welds, etc.) are similarly scored based upon the quality of the workmanship and the inspection technique.

100% inspection of all joints by industry-accepted practices warrants 2 points. Less than 100% or questionable or unknown inspection practice-reduces the points.

4. Backfill ....0–2 points

Type of backfill and procedures used ensured that no damage to the coating occurred. Uniform and (sometimes) compacted bedding material is necessary to properly support the pipe. Stress risers may arise from improper backfill or bedding material.

Knowledge and practice of good backfill/support techniques during construction warrants 2 points.

5. Handling ....0–2 points

Components, especially longer sections of line pipe, were handled in ways to minimize stresses. Cold-working of steel components for purposes of fit or line-up were minimized. Cold-working can cause high levels of residual stresses which in turn are a contributing factor to stress-corrosion phenomena. Handling includes storage of materials prior to installation. Protecting materials from harmful elements should be a part of proper handling.

The evaluator should award 2 points when he sees evidence of good material handling and storage techniques during and prior to construction.

6. Coating ....0–2 points

Coating application was supervised; coating was carefully inspected and repaired prior to final installation of pipe. This was a final step to ensure no last minute coating damage occurred during handling or final installation. Coating assessment is done in the *Corrosion Index* also, but at this stage, the human error potential is great. Proper handling and backfilling also directly impact the final condition of the coating. The best coating system can be easily sabotaged by simple errors in the final steps of installing the pipeline.

Again, the maximum points are awarded when the evaluator is satisfied that the constructors exercised due diligence in caring for the coating.

The evaluator must be careful in judging these items. Operators may have strong beliefs in how well these error-prevention activities were carried out, but may have little evidence to verify those beliefs. Evaluations of pipeline sections must reflect a consistency in awarding points and not be unduly influenced by unsubstantiated beliefs. A 'documentation-required' rule would help to ensure consistency.

Excavations, even years after initial installation, provide evidence of how well construction techniques were carried out. Findings such as damaged coatings, debris (temporary wood supports, weld rods, tools, rocks, etc.) buried with the pipeline, sloppy coatings over weld joints, etc., will still be present years later to indicate that perhaps not sufficient attention was paid during the construction process.

C. Operation   Suggested weighting ....35 pts
                             Prevention

Having considered design and construction, the third phase, operations, is perhaps the most critical from a human-error standpoint. This is a phase in which an error can produce an immediate failure.

Unlike the other phases, intervention opportunities here may be rare. The evaluator should look for a sense of professionalism in the way operations are conducted. A strong safety program is also evidence of attention being paid to error prevention. Both of these, professionalism and safety programs, are among the items believed to reduce errors.

The items in this section are somewhat redundant to each other, but are still thought to stand on their own merit. For example, better procedures enhance training; mechanical devices compliment training; better training and professionalism usually mean less supervision is required.

Operations is the stage where observability and controllability should be maximized. Wherever possible, intervention points should be established. These are steps in any process where actions contemplated or just completed are reviewed for correctness. At an intervention point, it is still possible to reverse the steps and place the system back in its prior (safe) condition.

A suggested point schedule to evaluate the operations phase is as follows:

1. Procedures .... 7 points
2. SCADA/Communications .... 5 points
3. Drug-testing .... 2 points
4. Safety programs .... 2 points
5. Surveys .... 2 points
6. Training .... 10 points
7. Mechanical Devices .... 7 points

1. Procedures ....0–7 points

The evaluator should satisfy himself that written procedures covering all aspects of pipeline operation exist. There is evidence that these procedures are actively used, reviewed, and revised. Such evidence might include filled-in checklists and copies of procedures in field locations or with field personnel. Ideally, use of procedures and checklists reduces variability. More consistent operations imply less opportunity for human error.

Examples of job procedures include:

- valve maintenance
- safety device inspection and calibration

- pipeline shutdown or start up
- pump operations
- product movement changes
- ROW maintenance
- flow meter calibrations
- instrument maintenance

The list goes on. Note that work near the line, but not actually involving the pipeline, is also included because such activities may affect the line. Unique or rare procedures should be developed and communicated with great care. A protocol should exist that covers these procedures—who develops them, who approves them, how is training done, how is compliance verified, how often are they reviewed. The non-routine is often the most dangerous.

The evaluator can check to see if procedures are in place for the most critical operations first: starting and stopping of major pieces of equipment, valve operations, changes in flow parameters, instruments taken out of service, etc.

A strong procedures program is an important part of reducing operational errors, as is seen by the point level. Maximum points should be awarded where procedure use is the highest. More is said about procedures in the training item.

2. SCADA/Communications ....0–5 points

Supervisory Control and Data Acquisition (SCADA) refers to the transmission of pipeline operational data (such as pressures, flows, temperatures, and product compositions) at sufficient points along the pipeline to allow monitoring of the line from a single location (Figure 6-3). In many cases, it also includes the transmission of data from the central monitoring location to points along the line to allow for remote operation of valves, pumps, etc. Devices called Remote Terminal Units (RTU) provide the interface between the pipeline data-gathering instruments and the conventional communication paths such as telephone lines, satellite transmission links, fiber optic cables, radio waves, or microwaves.

SCADA systems usually are designed to provide an overall view of the entire pipeline from one location. In so doing, system

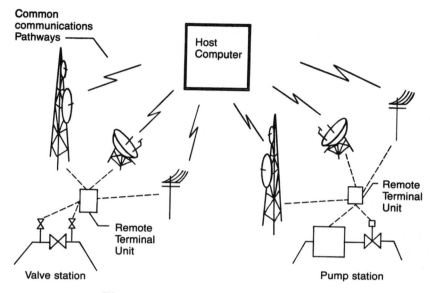

Common communications Pathways

Host Computer

Remote Terminal Unit

Remote Terminal Unit

Valve station

Pump station

Figure 6-3. Pipeline SCADA systems.

diagnosis, leak detection, transient analysis, and work coordination can be enhanced.

The main contribution of SCADA to human error avoidance is the fact that another set of eyes is watching pipeline operations and is hopefully consulted prior to field operations. A possible detractor is the possibility of errors emerging from the pipeline control center. More humans involved may imply more error potential, both from the field and from the control center. The emphasis should therefore be placed on how well the two locations are cooperating and cross-checking each other.

Protocol may specify the procedures in which both locations are involved. For example, the operating discipline could require communication between technicians in the field and the control center immediately before:

- valves opened or closed
- pumps and compressors started or stopped
- vendor flows started or stopped
- instruments taken out of service
- any maintenance that may affect the pipeline operation

Two-way communications between the field site and the control center should be a minimum condition to justify points in this section. Strictly for purposes of this section, a control center need not employ a SCADA system. The important aspect is that another source is consulted prior to any potentially upsetting actions. Telephone or radio communications, when properly applied, can be as effective as SCADA systems in preventing human error.

Five points should be awarded when the cross-checking is seen to be properly performed. Zero points should be awarded when there is no cross-checking being done.

3. Drug Testing ....0–2 points

Government regulations in the U.S. currently require drug testing programs for certain classes of employees in the transportation industry. The intent is to reduce the potential for human error due to an impairment of an individual. Company drug testing policies often include:

- random testing
- testing for cause
- pre-employment testing
- post-accident testing
- return-to-work testing

From a risk standpoint, finding and eliminating substance abuse in the pipeline work place reduces the potential for substance-abuse-related human errors.

A functioning drug testing program for pipeline employees who play substantial roles (DOT defines critical pipeline jobs) in pipeline operations should warrant 2 points.

4. Safety Programs ....0–2 points

A safety program is one of the nearly intangible factors in the risk equation. It is believed that a company-wide commitment to safety reduces the human error potential. Judging this level of commitment is difficult. At best we look for evidence of a commitment to satisfy. Such evidence may take the form of:

- written company statement of safety philosophy
- safety program designed with high level of employee participation—evidence of high participation is found
- strong safety performance record (recent history)
- housekeeping
- signs, slogans, etc., to show an environment tuned to safety
- full-time safety personnel

Most will agree that a company that promotes safety to this degree will have an impact on human error potential. A strong safety program should warrant 2 points.

5. Surveys ....0–2 points

While also covered in the risk indexes they specifically impact, surveys as a part of routine pipeline operations are again considered here. Examples of typical pipeline surveys include:

- close interval surveys
- coating condition surveys
- water crossing surveys
- deformation detection by pigging
- population density surveys
- depth of cover surveys
- sonar (subsea) surveys

Each is intended to identify areas of increased risk. A formal program of surveying, including proper documentation, implies a professional operation and a measure of risk reduction. Routine surveying indicates a more proactive, rather than reactive, approach to the operation. For the pipeline section being evaluated, points should be awarded based upon the number of surveys performed versus the number of useful surveys that could be performed there.

Maximum points should be awarded where surveying benefits are being optimized.

6. Training ....0–10 points

Training should be seen as the first line of defense against human error and accident reduction. For purposes of this risk assessment, training that concentrates on failure prevention is the most vital. The focus is on avoiding any failure of the pipeline system that may threaten life or property. This is in contrast to training that emphasizes protective equipment, first aid, injury prevention, and even emergency response. Such training is unquestionably critical, but its impact on the risk picture is indirect at best. This should be kept in mind as the training program is assessed for its contribution to risk reduction.

Obviously, different training is needed for different job functions and different experience levels. An effective training program, however, will have several key aspects, including common topics in which all pipeline employees should be trained. A point schedule can be developed to credit the program for each aspect that has been incorporated. An example (with detailed explanations afterwards) follows.

Documented Minimum Requirements .... 2 pts
Testing .... 2 pts
Topics covered:
    Product characteristics .... 0.5 pts
    Pipeline material stresses .... 0.5 pts
    Pipeline corrosion .... 0.5 pts
    Control and operations .... 0.5 pts
    Maintenance .... 0.5 pts
Emergency drills .... 0.5 pts
Job procedures (as appropriate) .... 2 pts
Scheduled re-training .... 1 pt

Documented Minimum Requirements. A document that specifically describes the body of knowledge which is expected of pipeline workers is the start of a good program. This document will ideally state the minimum knowledge requirements for each pipeline job position. Mastery of this body of knowledge will be verified before that position is worked by an employee. For example, a pump station operator will not be allowed to operate a station until he has demonstrated a command of all of the minimum requirements of that job. This should include station shutdowns, alarms, monitors, procedures, and the ability to recognize any abnormal conditions at the station.

Testing. A formal program should verify operator knowledge and identify deficiencies before they pose a threat to the pipeline system.

Tests that can be passed with less than 100% correctness may be failing to identify training weaknesses. Ideally, the operator should know exactly what knowledge he is expected to possess. The test should confirm that he does indeed possess this knowledge. He may be retested (within reasonable limits) until he has mastered the body of knowledge required for his job. Testing programs vary greatly in technique and effectiveness. It is left to the risk evaluator to satisfy himself that the testing achieves the desired results.

Topics Covered. Regardless of specific job, all pipeline operators (and arguably, all pipeline employees) should have some basic common knowledge. Some of these common areas may include:

- Product characteristics. Is the product transported flammable, toxic, reactive, carcinogenic? What are the safe exposure limits? If released does it form a cloud? Is the cloud heavier or lighter than air? Such knowledge decreases the chances of an operator making an incorrect decision due to ignorance of the product he is handling.
- Pipeline material stresses. How does the pipeline material react to stresses? What are indications of overstressing? What is the failure mode of the material? What is the weakest component in the system? Such basic knowledge must not be confused with engineering in the minds of the operators. All operators should understand these fundamental concepts to help avoid errors only—not to replace engineering decisions. With this knowledge though, an operator may find (and recognize the significance of) a bulge in the line indicating yielding had occurred. All trainees may gain a better appreciation of the consequences of a pipeline failure.
- Pipeline corrosion. As in the above topic, a basic understanding of pipeline corrosion and anti-corrosion systems may reduce the chances of errors. With such training, a field operator would be more alert to coating damage, the presence of other buried metal, or overhead power lines as potential threats to the pipeline. Office personnel may also have the opportunity to recognize a threat and bring it to the attention of the corrosion engineer, given a fundamental understanding of corrosion. A materials handler may spot a situation of incompatible metals that may have been overlooked in the design phase.

- Control and operations. This is most critical to the employees who actually perform the product movements, but all employees should understand how product is moved and controlled, at least in a general way. An operator who understands what manner of control is occurring upstream and downstream of his area of responsibility is less likely to make an error due to ignorance of the system. An engineer who understands the big picture of the pipeline system will anticipate all ramifications of changes to the system.
- Maintenance. A working knowledge of what is done and why it is being done may be valuable in preventing errors. A worker who knows how valves operate and why maintenance is necessary to their proper operation will be able to spot deficiencies in a related program or procedure. Inspection and calibration of instruments, especially safety devices, will usually be better done by a knowledgeable employee.

Emergency Drills. The role of emergency drills as a PROACTIVE risk reducer may be questioned. Emergency response in general is thought to play a role only after a failure has occurred and consequently is considered in the consequences portion of the *Leak Impact Factor.* Drills, however, may play a role in human error reduction as employees think through a simulated failure. The ensuing analysis and planning should lead to methods to further reduce the risk picture. The evaluator must decide what effect emergency drills have on the risk picture in a specific case.

Job Procedures. As required by specific employee duties, the greatest training emphasis should probably be placed on job procedures. The first step in avoiding improper actions of employees is to document the correct way to do things. Written and regularly reviewed procedures should cover all aspects of pipeline operation both in the field and in the control centers.

The use of procedures as a training tool is being measured here. Their use as an operational tool is covered in an earlier item.

Scheduled Retraining. Finally, experts agree that training is not permanent. Habits form, steps are bypassed, things are forgotten. Some manner of retraining and retesting is essential when relying upon a training program to reduce human error. The evaluator should satisfy himself that the retraining schedule is appropriate, and that the retesting adequately verifies employee skills.

7. Mechanical Error Preventers ....0–7 points

Sometimes dubbed "idiot-proofing," installing mechanical devices to prevent operator error is a proven effective risk reducer. Credit towards risk reduction should be given to any such device that impedes the accomplishment of an error. The premise here is that the operator is properly trained—the mechanical preventer serves to help avoid inattention errors. A simple padlock and chain can fit in this category, because such locks cause an operator to pause and hopefully consider action about to be taken. A more complex error-preventer is computer logic that will prevent certain actions from being performed out of sequence.

The point schedule for this category should reflect not only the effectiveness of the devices being rated, but also the possible consequences that are being prevented by the device. Judging this may need to be subjective, in the absence of good data. An example of a schedule with detailed explanations follows:

Three-way valves with dual instrumentation ....4 pts
Lock-out devices ....2 pts
Key-lock-sequence program ....2 pts
Computer permissives ....2 pts
Highlighting of critical instruments ....1 pts

In this schedule, points may be added for each application up to a maximum point value of 6 points. An application is valid only if the mechanical preventer is used in all instances of the scenario it is designed to prevent. If the section being evaluated has no possible applications, award the maximum points (7 points) because there is no potential for this type of human error.

Three-way Valves. It is common industry practice to install valves between instruments and pipeline components. The ability to isolate the instrument allows for maintenance of the instruments without taking the whole pipeline section out of service. Unfortunately, it also allows the opportunity for an instrument to be rendered useless if the isolating valve is left closed after the instrument maintenance is complete. Obviously if the instrument is a safety device such as a relief valve or pressure switch, it must not be isolated from the pipeline which it is protecting.

Three-way valves have one inlet and two outlets. By closing one outlet, the other is automatically opened. Hence, there is always an unobstructed outlet. When pressure switches, for instance, are installed at each outlet, one switch can be taken out of service and the other will always be operable. Both pressure switches cannot be simultaneously isolated. This is a prime example of a very effective mechanical preventer that reduces the possibility of a quite serious error. Points are awarded accordingly.

Lock-out Devices. These are most effective if they are not the norm. When an operator encounters a lock routinely, the attention-grabbing effect is lost. When the lock is an unusual feature, signifying unusual seriousness of the operation about to be undertaken, the operator is more likely to give the situation more serious attention.

Key-lock Sequence Programs. These are used primarily to avoid out-of-sequence type errors. If a job procedure calls for several operations to be performed in a certain sequence, and deviations from that prescribed sequence may cause serious problems, a key-lock sequence program may be employed to prevent any action from being taken prematurely. Such programs require an operator to use certain keys to unlock specific instruments or valves. Each key unlocks only a certain instrument and must then be used to get the next key. For instance, an operator uses his assigned key to unlock a panel of other keys. From this panel he can initially remove only key A. He uses key A to unlock and close valve X. When valve X is closed, key B becomes available to the operator. He uses key B to unlock and open valve Y. This makes key C available, etc. . . At the end of the sequence, he is able to remove key A and use it to retrieve his assigned key. These elaborate sequencing schemes involving operators and keys are being replaced by computer logic, but where they are used, they can be quite effective. It is important that the keys be non-defeatable to force operator adherence to the procedure.

Computer Permissives. These are the electronic equivalent to the key-locks described in the last section. By means of logic ladders, the computer prevents improper actions from being taken. A pump start command will not be executed if the valve line-up (proper up and downstream valves open or closed as required) is not correct. A command to open a valve will not execute if the pressure on either side of the valve is not in an acceptable range. Such electronic permissives are usually software programs that may reside in on-site or remotely located computers. A computer is not a minimum requirement, however, as simple solenoid switches or wiring arrangements may perform similar functions. The evaluator should satisfy himself that such permissives are adequate to perform the intended functions and that they are regularly tested and calibrated.

Highlighting of Critical Instruments. This is merely another method of bringing attention to critical operations. By painting a critical valve the color red, or tagging an instrument with a special designation, the operator will perhaps pause and consider his action again. Such pauses to reconsider may well prevent serious mistakes. Points should be awarded based upon how effective the evaluator deems the highlighting to be.

D. Maintenance   **Suggested weighting ....15 pts**
                                      Prevention

Improper maintenance is a type of error that can occur at several levels in the operation. Lack of management attention to maintenance, incorrect maintenance requirements or procedures, and mistakes made during the actual maintenance activities, are all errors that may directly or indirectly lead to a pipeline failure. The evaluator should again look for a sense of professionalism, as well as a high level of understanding of maintenance requirements for the equipment being used.

Note: This is an item that does not command a large share of the risk assessment points. However, many items in the overall pipeline risk assessment are dependent upon items in this section. A valve or instrument, which due to improper maintenance will not perform its intended function, negates any risk reduction that the device might have contributed. If the evaluator has concerns of proper operator actions in this area, he may need to adjust (downward) all maintenance-dependent items in the overall evaluation. Therefore, if this item scores low, it should serve as a trigger to initiate a re-evaluation of the pipeline.

Routine maintenance should include procedures and schedules for operating valves, inspecting cathodic protection equipment, testing/calibrating instrumentation and safety devices, corrosion inspections, painting, component replacement, lubrication of all moving parts, engine/pump/compressor maintenance, tank testing, etc.

Maintenance must also be done in a timely fashion. Maintenance frequency should be consistent with regulatory requirements and industry standards as a minimum.

The evaluator may wish to judge the strength of the maintenance program based upon the following items:

1. Documentation .... 2 points
2. Schedule .... 3 points
3. Procedures .... 10 points

The evaluator checks to see that:

1. Documentation ....0–2 points

A formal program of retaining all paperwork or databases dealing with all aspects of maintenance exists. This may include a file system or a computer database in active use. Any serious maintenance effort will have associated documentation. The ideal program will be constantly adjusting its maintenance practices based upon accurate data collection.

2. Schedule ....0–3 points

A formal schedule for routine maintenance based upon operating history, government regulations, and accepted industry practices exists. Again, this schedule will ideally reflect actual operating history, and within acceptable guidelines, be adjusted in response to that history.

3. Procedures ....0–10 points

Written procedures dealing with repairs and routine maintenance are readily available. Not only should these exist, it should also be clear that they are in active use by the maintenance personnel. Look for checklists, revision dates, and other evidence of their use. Procedures should help to ensure consistency. Specialized procedures are required to ensure that original design factors are still considered long after the designers are gone. A prime example is welding, where material changes such as hardness, fracture toughness, and corrosion resistance can be seriously affected by the welding process.

CHAPTER SEVEN

Leak Impact Factor

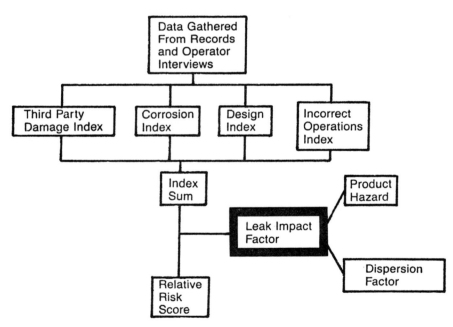

Figure 7-1. Components of the risk rating flowchart.

Leak Impact Factor

Leak Impact Factor = (Product + Hazard) ÷ (Dispersion Factor)
   A. Product Hazard
      (Acute+Chronic Hazards) .... 0–22 pts (p. 167)

1. Acute Hazards
   a. N_f ....   0–4 (p. 168)
   b. N_r ....   0–4 (p. 169)
   c. N_h .... $\underline{0\text{–}4}$ (p. 171)

   Total (N_h + N_r + N_f) 0–12
2. Chronic Hazard, RQ ....   0–10 (p. 172)
B. Dispersion Factor (Spill Score)
   ÷ (Population Score) ....   0–6 (p. 181)
   1. Liquid Spill or Vapor Spill ....   0–6 (p. 183)
   2. Population Density ....   0–4 (p. 203)

Note: The *Leak Impact Factor* will be used to adjust the index scores to reflect the consequences of a failure. A higher point score for the *Leak Impact Factor* represents a higher risk.

Leak Impact Factor

Up to this point, possible pipeline failure initiators have been assessed. These initiators define *What can go wrong.* As part of the assessment, actions or devices that are designed to prevent these failure initiators were also considered. These preventions affect the *How likely is it?* follow-up question to *What can go wrong?*

The last portion of the risk assessment addresses the question of *What are the consequences?* The complete picture then becomes:

- the potential hazards
- the probabilities of the hazards occurring
- the consequences of the hazards occurring

The consequence factor begins at the point of pipeline failure. The title of this chapter, *Leak Impact Factor (LIF)*, emphasizes this. What is the impact of a pipeline leak? The answer depends upon two factors:

- the pipeline product
- the pipeline surroundings

Unfortunately, the interaction between these two factors can be immensely complex and almost impossible to model. The possible leak rates, weather conditions, soil types, populations nearby, etc. are in themselves highly variable and unpredictable. When the interactions between these and the product characteristics are also considered, the problem becomes solvable only through assumptions and approximations.

The *Leak Impact Factor* is calculated from an analysis of the product characteristics and the spill or release characteristics. While simplifying assumptions are used, enough distinctions are made to ensure meaningful risk assessments.

In studying the impact of a leak, we make a distinction between acute and chronic hazards. Acute can mean sudden onset, or demanding urgent attention, or of short duration. Hazards such as fire, explosion, or contact toxicity are considered to be acute hazards. They are immediate threats caused by a leak.

Chronic means marked by a long duration. A time variable is therefore implied. Hazards such as groundwater contamination, carcinogenicity, and other long-term health effects are considered to be chronic hazards. Many releases that can cause damage to the environment are chronic hazards—they can cause long-term effects and have the potential to worsen with the passage of time.

The primary difference between acute and chronic hazards is the time factor. The immediate hazard, which forms instantly upon initiation of the event, grows to its highest point within a few minutes and then reduces, is an acute hazard. The hazard that potentially grows worse with the passage of time is a chronic hazard.

For example, a natural gas release poses mostly an acute hazard. The largest possible gas cloud normally forms immediately, creating a fire/explosion hazard, and then begins to shrink as pipeline pressure decreases. If the cloud does not find an ignition source, the hazard is reduced as the vapor cloud shrinks. (If the natural gas vapors can accumulate inside a building, the hazard may become more severe as time passes—it then becomes a chronic hazard.)

The spill of gasoline or kerosene may be more chronic in nature. As time passes, the dispersion of spilled product increases, extending the area of opportunity for ignition and increasing the area that is at risk should ignition occur. The environmental harm is also more widespread as the leak continues.

The evaluator should imagine where his product would fit on a scale such as that shown in Figure 7-2. Some product hazards are almost purely acute in nature, such as natural gas. These are shown on the

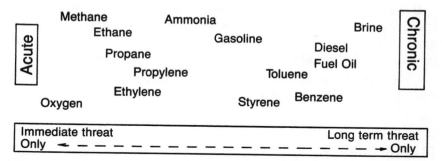

Figure 7-2. Relative acute-chronic hazard scale for some pipeline products.

left edge of the scale. Others, such as brine, pose no immediate (acute) threat, but cause environmental harm as a chronic hazard. These appear on the far right of the scale.

Product Characteristics

The primary factor in determining the nature of the hazard is the pipeline product itself.

Most products will have some acute hazard characteristics and some chronic hazard characteristics. The evaluator should fix in his mind which of these effects is most critical. Referring again to Figure 7-2, a hypothetical scale can be used to illustrate where some common pipeline products may fit in relation to each other. A product's location on this scale is dependent upon how readily it disperses (the persistence), how much long-term hazard, and how much short-term hazard it presents.

Acute Hazards

Regardless of whether it is a gas or a liquid, the product must be assessed in terms of its toxicity, flammability, and reactivity. These are the acute hazards. One industry-accepted scale for rating product hazards comes from the National Fire Prevention Association (NFPA). This scale rates materials based on the threat to emergency response personnel (acute hazards). The potential threat is examined in terms of flammability, toxicity, and reactivity.

If the product is a mixture of several components, the mixture itself could be rated. However, an alternative would be to base the assessment on the most hazardous component, since NFPA data might be more readily available for the components individually.

Unlike the previous point scoring systems described in this book, the *Leak Impact Factor* reflects increasing hazard with increasing point values.

Flammability, N_f

Many common pipeline products are very flammable. The greatest hazard from most hydrocarbons is from flammability.

N_f is the symbol used to designate the flammability rating of a substance according to the NFPA scale. The five-point scale shows, in a relative way, how susceptible the product is to combustion. The flash point is one indicator of this flammability. Flash point is defined as the minimum temperature at which the vapor over a flammable liquid will 'flash' when exposed to a free flame. It tells us what temperature is required to release enough flammable vapors to support a flame. Materials with a low flash point ($< 100°F$) ignite and burn readily and are deemed to be flammable. If this material also has a boiling point less than $100°F$, it is considered to be in the most flammable class. This includes methane, propane, ethylene, and ethane. The next highest class of substances has flash points less than $100°F$ and boiling points greater than $100°F$. In this class, less product vaporizes and forms flammable mixtures with the air. It includes gasoline, crude petroleum, naphtha, and certain jet fuels.

A material is termed combustible if its flash point is greater than $100°F$, and it will still burn. This class includes diesel and kerosene. Examples of non-combustibles are bromine and chlorine.

Use the following table or Appendix A to determine the NFPA N_f value (FP = Flash Point, BP = Boiling Point [9]):

Non-combustible .... $N_f = 0$
FP > 200°F .... $N_f = 1$
100°F < FP < 200°F... $N_f = 2$
FP < 100°F and BP < 100°F .... $N_f = 3$
FP < 73°F and BP < 100°F.................................. $N_f = 4$

More will be said about flammability in the discussion of vapor cloud dispersion later in this chapter.

Reactivity, N_r

Occasionally, a pipeline will transport a material that is unstable under certain conditions. A reaction with air, water, or with itself, could be potentially dangerous. To account for this possible increase in hazard, a reactivity rating should be included in the assessment of the product. The NFPA value, N_r, is used to do this.

While a good beginning point, the N_r value should be modified when the pipeline operator has evidence that the substance is more reactive than the rating implies. An example of this might be ethylene. A rather common chain of events in pipeline operations can initiate a destructive series of detonations inside the line. This is a type of reactivity that should indicate to the handler that ethylene is unstable under certain conditions and presents an increased risk due to that instability. The published N_r value of 2 might not adequately cover this special hazard for ethylene in pipelines.

Use the following table or Appendix A to determine the N_r value [9].

N_r = 0 Substance is completely stable, even when heated
    under fire conditions.
N_r = 1 Mild reactivity upon heating with pressure.
N_r = 2 Significant reactivity, even without heating.
N_r = 3 Detonation possible with confinement.
N_r = 4 Detonation possible without confinement.

Note that reactivity includes self-reactivity (instability) and reactivity with water.

The reactivity value (Nr) can be obtained more objectively by using the peak temperature of the lowest exotherm value as follows [9]:

Exotherm, °C	N_r
>400	0
305–400	1
215–305	2
125–215	3
<125	4

The immediate threat from the potential energy of a pressurized pipeline is also considered here. This threat includes debris and pipe fragments that could become projectiles in the event of a catastrophic pipeline failure. Accounting for internal pressure in this item quantifies the intuitive belief that a pressurized container poses a threat that is not present in a non-pressurized container.

The increased hazard due solely to the internal pressure is thought to be rather small because the danger zone is usually limited for a buried line. When the evaluator sees an increased threat, such as an aboveground section in a populated area, he may wish to adjust the reactivity rating upward in point value.

In general, a compressed gas will have the greater potential energy and hence the greater chance to do damage. This is in comparison to an incompressible fluid.

The pressure hazard is directly proportional to the amount of internal pressure in the line. While the MAOP could be used here, this would not differentiate between the upstream sections (often higher pressures) and the downstream sections (usually lower pressures). One approach would be to create a hypothetical pressure profile of the entire line and, from this, identify normal maximum pressures in the section being evaluated. Using these pressures, points can be assessed to reflect the risk due to pressure.

So, to the N_r value determined above, a pressure factor can be added as follows:

Incompressible fluids (liquids)   Pressure Factor
  0–100 psig internal pressure .... 0 points
  >100 psig .... 1 point

Compressible fluids (gases)
  0–50 psig .... 0 points
  51–200 psig .... 1 point
  >200 psig .... 2 points

Total point values for N_r should not be increased beyond 4 points, however, because that would minimize the impact of the flammability and toxicity factors, N_f and N_h, whose maximum point scores are 4 points.

Example:

A natural gas pipeline is being evaluated. In this particular section, the normal maximum pressure is 500 psig. The evaluator determines from Appendix A that the N_r for methane is 0. To this, he adds 2 points to account for the high pressure of this compressible fluid. Total score for reactivity is therefore 2 points.

Toxicity, N_h

The NFPA rating for a material's health factor is N_h. The N_h value only considers the health hazard in terms of how that hazard complicates the response of emergency personnel. Long-term exposure effects must be assessed using an additional scale. Long-term health effects will be covered in the assessment of chronic hazards associated with product spills.

As defined in NFPA 704, the toxicity of the pipeline product is scored on the following scale [9]:

N_h = 0 No hazard beyond that of ordinary combustibles.
N_h = 1 Only minor residual injury is likely.
N_h = 2 Prompt medical attention required to avoid temporary incapacitation.
N_h = 3 Materials causing serious temporary or residual injury.
N_h = 4 Short exposure causes death or major injury.

Appendix A lists the N_h value for many substances commonly transported by pipeline.

Acute Hazard

The Acute Hazard is now obtained by adding the scores as follows:

$(N_f + N_r + N_h)$ = Acute Hazard (0–12 points)

A score of 12 points represents a substance that poses the most severe hazard in all three of the characteristics studied.

The possible point values are low, but this is a multiplying factor. As such, it will have a substantial effect on the total risk score.

Few preventive actions are able to substantially reduce acute hazards. To be effective, a preventive action would have to change the characteristics of the hazard itself. Quenching a vapor release instantly or otherwise preventing the formation of a hazardous cloud would be one example of how the hazard could be changed. While the probability and the consequences of the hazardous event can certainly be managed, the state-of-the-art is not thought to be so advanced as to change the acute hazard of a substance as it is being released.

Environmental/Chronic Hazard, RQ

A very serious threat from a pipeline is the potential loss of life caused by a release of the pipeline contents. This is usually considered to be an acute, immediate threat. Another quite serious threat that may also ultimately lead to loss of life is the contamination of the environment due to the release of the pipeline contents. Though not usually as immediate a threat as toxicity or flammability, environmental contamination ultimately affects life, with possible far-reaching consequences.

This section offers a method to rate those consequences that are of a more chronic nature. We will build upon the previous section in doing this. From the acute leak impact consequences model, we can rank the hazard from fire and explosion for the flammables and from direct contact for the toxic materials. These hazards were analyzed as short-term threats only. We are now ready to examine the longer-term hazards associated with pipeline releases.

Figure 7-3 illustrates how the chronic hazard associated with pipeline spills will be addressed.

The first criterion is whether or not the pipeline product is considered to be hazardous. For this determination, the U.S. government regulations will be used. The regulations loosely define a hazardous substance as a substance that can potentially cause harm to humans or to the environment. Hazardous substances are more specifically defined in a variety of regulations including the Clean Water Act (CWA), the Clean Air Act (CAA), the Resource Conservation and Recovery Act (RCRA), and the Comprehensive Environmental Response, Compensation and Liability Act (CERCLA, also known as Superfund). If the pipeline product is considered by any of these sources to be hazardous, a reportable spill quantity (RQ) category designation is assigned under CERCLA (Figure 7-3). These RQ designations will be used in our pipeline risk assessment to help rate hazardous products from a chronic standpoint.

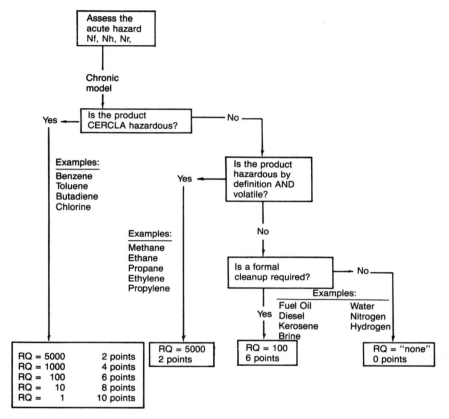

Figure 7-3. Determination of RQ.

The more hazardous substances have smaller reportable spill quantities. Larger amounts of more benign substances may be spilled before the environment is damaged. Less hazardous substances, therefore, have larger reportable spill quantities. The designations are categories X, A, B, C, and D, corresponding to spill quantities of 1, 10, 100, 1,000, and 5,000 pounds, respectively. Class X, a 1-pound spill, is the category for substances posing the most serious threat. Class D, a 5,000-pound spill, is the category for the least harmful substances.

The EPA clearly states that its RQ designations are not created as agency judgments of the degree of hazard of specific chemical spills. That is, the system is NOT intended to say that a 9-pound spill of a class A substance is not a problem, while a 10-pound spill is. The RQ is designed to be a trigger point at which the government can investigate a spill to assess the hazards and to gauge its response to the spill. The criteria used in determining the RQ are, however, appropriate for our purposes in ranking the relative environmental hazards of spills.

Classifying a chemical into one of these Reportable Quantities categories is a non-trivial exercise outlined in U.S. Regulations, 40 CFR Parts 117 and 302. The primary criteria considered include aquatic toxicity, mammalian toxicity (oral, dermal, inhalation), ignitability, reactivity, chronic toxicity, and potential carcinogenicity. The lowest of these criteria (the worst case) will determine the initial RQ of the chemical.

The initial RQ may then be adjusted by analysis of the secondary criteria of biodegradation, hydrolysis, and photolysis. These secondary characteristics provide evidence as to how quickly the chemical can be safely assimilated into the environment. A chemical that is quickly converted into harmless compounds poses less risk to the environment. So-called "persistent" chemicals receive higher hazard ratings.

The CERCLA Reportable Quantity list has been revised since its inception, and will probably be continually revised. One weakness of the system is that the best available knowledge may not always be included in the most current version. An operator who is intimately familiar with a substance MAY be in a better position to rate that product relative to some others. When operator experience suggests that the substance is worse than the published CERCLA RQ implies, the evaluator should probably revise the number to a more severe rating. This can be done with the understanding that the CERCLA rating is subject to periodic review and will most likely be updated as better information becomes available. If the operator, on the other hand, feels that the substance is being rated too severely, the evaluator should recognize that the operator may not realize all aspects of the risk. It is recommended that RQ ratings should NOT be reduced in severity rating based solely upon operator opinions.

Using the RQ factor incorporates some redundancy to the already assigned NFPA ratings for acute hazards. However, the overlap is not complete. The RQ factor adds information on chronic toxicity, carcinogenicity, persistence, and toxicity to non-humans; none of which are included in the NFPA ratings. The overlap does specifically occur in acute toxicity, flammability, and reactivity. This causes no problems for a relative risk analysis.

Primary Criteria (See Briggum et al., pp. F-13, 14 [4].)

The following is a brief summary of each of the CERCLA primary criteria.

1. Aquatic Toxicity. Originally developed under the Clean Water Act, the scale for aquatic toxicity is based on LC_{50}, the concentration of chemical that is lethal to one-half of the test population of aquatic animals upon continuous exposure for 96 hours.

Table 7-1
Aquatic Toxicity

RQ	Aquatic Toxicity (LC_{50} range)
1	<0.1 mg/L
10	0.1 to 1.0 mg/L
100	1 to 10 mg/L
1,000	10 to 100 mg/L
5,000	100 to 500 mg/L

2. Mammalian Toxicity. This is a five level scale for oral, dermal, and inhalation toxicity for mammals. It is based upon LC_{50} data as well as LD_{50} (the dose required to cause the death of 50% of the test population) data and is shown below.

Table 7-2
Mammalian Toxicity

RQ lbs	Mammalian Toxicity (oral LD_{50} range)	Mammalian Toxicity (dermal LD_{50} range)	Mammalian Toxicity (inhalation LC_{50} range)
1	<0.1 mg/kg	<0.04 mg/kg	<0.4 ppm
10	0.1 to 1	0.04 to 0.4	0.4 to 4
100	1 to 10	0.4 to 4	4 to 40
1000	10 to 100	4 to 40	40 to 400
5000	100 to 500	40 to 200	400 to 2000

See *Notes on Toxicity* later in this chapter.

3. Ignitability and Reactivity. Ignitability is based upon flash point and boiling point in the same fashion as the acute characteristic,

N_f. Reactivity is based on a substance's reactivity with water and with itself. For our purposes, it also includes pressure effects in the assessment of acute hazards.

4. Chronic toxicity. To evaluate the toxicity, a scoring methodology assigns values based on the minimum effective dose for repeated exposures and the severity of the effects caused by exposure. This scoring is a function of prolonged exposure, as opposed to the acute factor, N_h, which deals with short-term exposure only. The score determination methodology is found in U.S. regulations (48 CFR 23564).

5. Potential Carcinogenicity. This scoring is based upon a high 'weight of evidence' designation (either a "known," "probable," or "possible" human carcinogen) coupled with a potency rating. The potency rating reflects the relative strength of a substance to elicit a carcinogenic response. The net result is a high, medium, or low hazard ranking which corresponds to RQs of 1, 10, and 100 pounds, respectively [12].

Secondary Criteria. As previously stated, the final RQ rating may be adjusted by evaluating the persistence of the substance in the environment. The susceptibility to biodegradation, hydrolysis, and photolysis allows certain substances to have their RQ ratings raised one category. To be considered for the upgrade, the substance has to pass initial criteria dealing with tendency to bioaccumulate, environmental persistence, presence of unusual hazards (such as high reactivity), and the existence of hazardous degradation or transformation products. If the substance is not excluded because of these items, it may be adjusted upwards one RQ category if it shows a very low persistence.

Unfortunately, petroleum, petroleum feedstocks, natural gas, crude oil, and refined petroleum products are specifically excluded from the EPA's reportable quantity requirements under CERCLA. Because these products comprise a high percentage of substances transported by pipeline, an alternative scoring system must be used. This requires a deviation from the direct application of the EPA rating system when petroleum products are evaluated. We can, however, extend the spirit of the EPA system to encompass all common pipeline products, for our purposes here. This is done by assigning RQ equivalent classifications to substances that are not assigned an RQ classification by the EPA.

For the products not specifically listed as hazardous by EPA regulatory agencies, a general definition is offered. If the any one of the following four properties are present, the substance is considered to be hazardous [4].

1. Ignitability. Defined as a liquid with a flash point less than 60°C or a non-liquid that can spontaneously cause a fire through friction, absorption of moisture, or spontaneous chemical changes and will burn vigorously and persistently.
2. Corrosivity. Defined as liquids with pH $\leq$ 2 or $\geq$ 12.5, or with the ability to corrode steel at a rate of 6.35 millimeters per year at 55°C.
3. Reactivity. Defined as a substance that is normally unstable, reacts violently with water, forms potentially violent mixtures with water, generates toxic fumes when mixed with water, is capable of detonation or explosion, or is classified as an explosive under DOT regulations.
4. Extraction procedure toxicity. This is defined by a special test procedure that looks for concentrations of materials listed as contaminants in the Safe Drinking Water Act's list of National Interim Primary Drinking Water Regulation contaminants. (See Briggum et al., pp 46–47 [4].)

Although the petroleum products are specifically excluded from regulatory control, these definitions would obviously include most pipeline hydrocarbon products. This then becomes the second criterion to be made in the evaluation of pipeline products:

Products that are not specifically listed with an EPA assigned RQ but **do** fit the definition of hazardous are now divided into categories of volatile or non-volatile. Products that are not specifically listed and that do **not** meet the definition of hazardous set forth above, are assumed to have a RQ designation of "none." (See Figure 7-3.)

Following the hazardous branch of the flowchart (Figure 7-3), we now assess the volatile substances. Highly volatile hazardous products of concern produce vapors, which when released into the atmosphere, cause potential acute hazards, but usually only minimal chronic hazards. Common pipeline products that will fall into this category include methane, ethane, propane, ethylene, propylene, and other liquified petroleum gases.

We can assume that the bulk of the hazard from highly volatile substances occurs in leaks to the atmosphere. We assume that all leaks of such products into any of the three possible environmental media (air, soil, water) will ultimately cause a release to the air. We can then surmise that the hazard from these highly volatile liquids is mostly addressed in the atmospheric dispersion modelling analysis that will be performed in the acute leak impact consequences analysis. The chronic part of this leak scenario is thought to be in the potential for 1) residual hydrocarbons to be trapped in soil or buildings, and pose a later flammability threat, and 2) the so-called "greenhouse" gases that are thought to be harmful to the ozone layer of the atmosphere. These threats warrant an $RQ_{equivalent}$ of 5,000 pounds in this ranking system.

This leaves the less-volatile hazardous substances. Included here are petroleum products such as kerosene, jet fuel, gasoline, diesel oil, and crude oils. For spills of these substances, the acute hazards are already addressed in the flammability, toxicity and reactivity assessment. Now, the chronic effects such as pollution of surface waters or groundwater, and soil contamination are taken into account.

Spills of non-volatile substances must be assessed as much from an environmental insult basis as from an acute hazard basis. This in no way minimizes the hazard from flammability, however. The acute threat from spilled flammable liquids is addressed in the acute portion of the leak impact. The longer-term impact of spilled petroleum products is obtained by assigning an RQ number to these spills. It is recommended that these products be classified as category B spills (Reportable quantities of 100 pound) unless strong evidence places them in another category. This means the $RQ_{equivalent}$ is 100 pounds. An example of evidence sufficient to move the product down one category (more hazardous) would be the presence of a significant amount of category X or category A material present (such as methylene chloride—category X). This is discussed further below. Evidence that could move the petroleum product into a category C or category D (less hazardous) would be high volatility or high biodegradation rates.

To make further distinctions within this group involves more complex determinations. The value of these additional determinations is not thought to outweigh the additional costs. For instance, it can perhaps be generally stated that the heavier petroleum products will biodegrade at a slower rate than the lighter substances. This is because the degradability is linked to the solubility and the lighter products are usually more soluble. However, it can also be generally stated that the lighter petroleum substances may more easily penetrate the soil and reach deeper groundwater regions. This is also a solubility phenomenon. We now have conflicting results of a single property. To adequately include the property of density (or solubility), we would have to balance the benefits of quicker degradation with the potential of more widespread environmental harm.

We have now established a methodology to assign a ranking, in the form of an RQ category, for each pipeline product. An important exception to the general methodology is noted. If the quantity spilled is great enough to trigger an RQ of some trace component, this RQ should govern. This scenario may occur often because we are using complete line rupture as the main leak quantity determinant. For example, a crude oil product that has 1% benzene would reach the benzene RQ number on any spill greater than 1,000 pounds. This is because the benzene RQ is 10 pounds and 1% of 1,000-pound spill of product containing 1% benzene means that 10 pounds of benzene was spilled.

To easily account for this general exception to the RQ assignment, the evaluator should start with the leak quantity calculation. He can then work from the CERCLA list and determine the maximum percentage for each trace component that must be present in the product stream before that component governs the RQ determination. Comparing this to an actual product analysis will point out the worst case component that will determine the final RQ rating. An example will illustrate this.

Example

An 8-in. pipeline that transports a gasoline that is known to contain the CERCLA hazardous substances benzene, toluene, and xylene is being evaluated. The leak quantity is calculated from the line size and the normal operating pressure (normal pressures instead of maximum allowable pressures are used throughout this company's evaluations) to be 10,000 lbs.

This calculated leak quantity, 10,000 pounds, is now used to determine component percentages that will trigger their respective RQs for this spill.

benzene (RQ = 10) ....*10/10000 = 0.001 = 0.1%*
toluene (RQ = 1000) .... *1000/10000 = 0.1 = 10%*
xylene (RQ = 1000) .... *1000/10000 = 0.1 = 10%*

The evaluator can now look at an actual analysis to see if the actual product stream exceeds any of these weight percentages.

If the benzene concentration is less than 0.1% and the toluene and xylene concentrations are each less than 10%, then the RQ is set at 100 pounds, the default value for gasoline. If, however, actual analysis shows the benzene concentration to be 0.7%, then the benzene RQ set at 10 pounds governs. This is because, more than 10 pounds of benzene will be spilled in a 10,000 pound spill of this particular gasoline stream.

Gasolines generally are rich in benzene, but they are also fairly volatile. Heating oils, diesel, and kerosene are more persistent, but may contain less toxicants and suspected carcinogens. Crude oils, of course, cover a wide range of viscosities and compositions. The pipeline operator will no doubt be familiar with his products and their properties.

Note that there is a 2-point spread between each RQ classification. The evaluator may pick the midpoint between two RQs if he has special information that makes it difficult to strictly follow the suggested scoring. Once again, he must be consistent in his scoring.

Notes on Toxicity

An important part of the degree of consequences, both acute and chronic, is toxicity. The following provides further discussion on toxicity.

The degree of toxic hazard is usually expressed in terms of exposure limits to humans. Exposure is only an estimate of the more meaningful measure which is dosage. The dose is the amount of the product that gets into the human body. Health experts have established dosage limits beyond which permanent damage to the human may occur. Because the intake (dose) is a quantity that is difficult to measure, it is estimated by measuring the opportunity for intaking a given dose. This intake estimate is the exposure.

There are three recognized exposure pathways; inhalation, ingestion, and dermal contact. Breathing contaminated air, eating contaminated foods, or coming into skin contact with the contaminant can all lead to the increased dose level within the body. Some of the exposure pathways can extend for long distances, over long periods of time from the point of contaminant release. Plants and animals that absorb the contaminant may reach humans only after several levels of the food chain. Groundwater contamination may spread over great distances and remain undetected for long periods. Calculations are performed to estimate dosages for each exposure pathway.

EPA Ingestion Route calculations include approximate consumption rates for drinking water, fruits and vegetables, beef and dairy products, fish and shellfish, and soil ingestion (by children). These consumption rates, based upon age and sex of population affected, are multiplied by the contaminant concentration and by the exposure duration. This value, divided by the body weight and lifespan, yields the lifetime average ingestion exposure.

In a similar calculation, the lifetime average inhalation exposure yields an estimate of the inhalation route exposure. This is based upon studies of movement of gases into and out of the lungs (pulmonary ventilation). The calculation includes considerations for activity levels, age, and sex.

The dermal route dose is obtained by estimating the dermal exposure and then adjusting for the absorption of the contaminant. Included in this determination are estimates of body surface area (which in turn is dependent upon age and sex), and typical clothing of the exposed population.

In each of these determinations, estimates are made of activity times in outdoor play/work, showering, driving, etc. Lifespans are similarly estimated for the population under study.

It is not proposed that all these parameters be individually estimated for purposes of this risk assessment. The evaluator should realize the simplifications he is making, however, in rating spills here. Because we are only concerned with relative hazards, accuracy is not lost, but absolute risk determination is not possible without the more formal methods.

Dispersion

As modelled by physics and thermodynamics, spilled product will always seek a lower energy state. The laws of entropy tell us that the system will become increasingly disordered. The product will mix and intersperse itself with its new environment in a non-reversible process.

The spill has introduced a stress into the system. The system will react to relieve the stress by spreading the new energy throughout the system until a new equilibrium is established.

For purposes of this assessment, accurate modelling of the dispersion of spilled product is not necessary. It is the propensity to do harm which is of interest. A substance that causes great damage even at low concentrations, released into an area that allows rapid and wide-ranging spreading, is the greatest hazard.

If a product escapes from the pipeline, it is released as either a gas or a liquid (or a combination of the two). As a gas, the product has more degrees of freedom and will disperse more readily. This may be bad or good, since the product may cover more area, but in a less concentrated form. A flammable gas will entrain oxygen as it disperses, hence becoming an ignitable mixture. A toxic gas may quickly be reduced to safe exposure levels as its concentration decreases.

The relative density of the gas in the atmosphere will partly determine its dispersion characteristics. A heavier gas will generally stay more concentrated and accumulate in low lying areas. A lighter gas should rise due to its buoyancy in the air. Every density of gas will be affected to some extent by air temperature, wind currents, and terrain.

A product that stays in liquid form when released from the pipeline poses different problems. Environmental insult, including ground water contamination, and flammability are the most immediate problems, although toxicity can play a role in both the short and long-term scenario.

We must always remember the sensitivity of the environment to certain substances. Contaminations in the few parts per billion or even parts per trillion are often of concern. If contamination is defined as 10 parts per billion, a 10-gallon spill of a solvent can contaminate a billion gallons of groundwater. A 5,000-gallon spill from a pipeline can contaminate 500 billion gallons of groundwater to 10 ppb. The potential contamination is determined by the simple formula:

$$V_1 \times C_1 = V_{gw} \times C_{gw}$$

where V_1 = volume of spill
V_{gw} = volume of groundwater contaminated
C_1 = average concentration of contaminant in spilled material
C_{gw} = average concentration of contaminant in groundwater

Reductions in the harmful properties of the substance reduce the hazard. This may occur through natural processes such as biodegradation, photolysis, and hydrolysis. If the byproducts of these reactions are less harmful than the original substance, which they often are, the hazard is proportionally reduced. Reductions in the range of dispersion of the substance also reduce the hazard. From a risk standpoint, the degree of dispersion impacts the area of opportunity because more wide-ranging contamination offers greater chances to harm life. Dispersion is addressed in the Spill Score.

The Dispersion Factor will be calculated from an analysis of the spill itself and the population near to the pipeline.

Determining the Spill Scores

To assess a spill score, the evaluator must first determine which state, vapor or liquid, will be present after a pipeline failure. If both states exist, the more severe hazard should govern.

Vapor Clouds (Vapor Spills)

Of great interest to risk evaluators are the characteristics of vapor cloud formation and dispersion following a pipeline release. Vapor can be formed from product that is initially in a gaseous state or from a product that vaporizes as it escapes or as it accumulates in pools on the ground. The amount of vapor put into the air, and the vapor concentrations at varying distances from the source are the subject of many modelling efforts.

A reasonable question might well be: *How does vapor cloud formation impact the risk picture?* Two potential hazards are created by a vapor cloud. One hazard occurs if the product in the cloud is toxic. The threat is then to any susceptible life forms that come into contact with the cloud. Larger clouds provide a greater area of opportunity for this contact to occur and hence carry a greater hazard. Note that a cloud may also be toxic in that it displaces oxygen and suffocates the life form.

The second hazard occurs if the cloud is flammable. The threat then is that the cloud will find an ignition source, causing fire and/or explosion. Larger clouds have a greater chance of finding an ignition source and also increase the damage potential because more flammable material is involved. Of course, the vapor can contain both hazards: toxicity and flammability. The cloud may not be visible; it may be a very low concentration cloud—only a few parts per million of released product in air. Even at these low concentrations, however, a cloud exists and can be a hazard.

Again, a vapor cloud that covers more ground, either due to its size or its cohesiveness, has a greater area of opportunity to find an ignition source or to harm living creatures. At the other extreme, a pipeline product that does not form a vapor cloud at all, does not create the same kind of hazard as a material which does vaporize. Hazards from the release of liquids are more chronic in nature and are covered later in the spill score.

The hazard is being examined in parts independent from one another. The cloud characteristics are examined on their own, without thought as to whether a cloud is "bad" or not. The "badness" of the cloud is a result of the product acute hazards: N_f, N_h, N_r. Together, these two parts define the acute hazard.

When an escaping pipeline product forms a vapor cloud, the entire range of possible concentrations of the product-air mixture exist. At some point, the fuel-to-air ratio will be in the flammable range. A flammable gas will therefore be ignitable at this point in the cloud. Although ignition is not necessarily inevitable, there is often a high probability due to the large number of possible ignition sources—cigarettes, engines, open flames, residential heaters, sparks, just to name a few. It is conservative to assume, then, that an ignition source will come into contact with the proper fuel-to-air ratio at some point during the release. The consequences of this contact range from a simple fire to a massive detonation and fireball.

The manner in which a vapor cloud goes from burning to exploding is not well understood. Upon ignition, a flame propagates through the cloud, entraining surrounding air and fuel from the cloud. If the flame propagation speed becomes high enough, a fireball and possibly a detonation can occur. The fireball can radiate damaging heat far beyond the actual flame boundaries, causing skin and eye damage and secondary fires. If the cloud is large enough, a 'fire storm' can be created, generating its own winds, causing far reaching secondary fires and radiant heat damage.

Should a detonation occur, even more widespread damage is possible. A detonation can generate powerful blast waves also, far beyond the actual cloud boundaries. As a matter of fact, most hydrocarbon-air mixtures have heats of combustion greater than the heat of explosion of TNT [1]. The possibility of vapor cloud explosions is enhanced by closed areas, including partial enclosures created by trees or buildings. Unconfined vapor cloud explosions are nonetheless a real danger. Certain military bombs are designed to take advantage of the increased blast potential created by the ignition of an unconfined cloud of hydrocarbon-air vapor.

Many variables affect the dispersion of vapor clouds. The extreme complexities make the problem only approximately solvable for even a relatively closed system. An example of a somewhat closed system is a well-defined leak from a small chemical plant where the terrain is known and constant and where weather conditions can be reasonably estimated from real time data. A cross-country pipeline, on the other hand, complicates the problem by adding variables such as soil conditions (moisture content, temperature, heat transfer rates, etc.), often constantly changing terrain and weather patterns (amount of sunshine, wind speed and direction, humidity, elevation, etc.), and even the difficulties in locating the source of the leak.

Even though it vaporizes quickly, a highly volatile pipeline product can form a liquid pool immediately after release. This could be the case with products such as propane or ethylene. The pool would then become a secondary source of the vapors. Vapor generation would be dictated by the temperature of the pool surface which in turn is controlled by the air temperature, the wind speed over the pool, the amount of sunshine to reach the pool, and the heat transfer from the soil (Figure 7-4). The soil heat transfer is in turn governed by soil moisture content, soil type, and the weather of the previous few weeks! Even if all these factors could be accurately measured, the system is still a non-linear relationship that cannot be exactly solved.

Ideally, the risk evaluator would use data from the dispersion modelling of specific pipeline releases under specific conditions. From this data, he could select worst case scenarios for each pipeline section. From these scenarios, he could then develop a schedule to rank the relative dispersions predicted. A prediction of a large, fairly cohesive cloud would carry a higher hazard value than a smaller, more readily dispersed cloud.

Unfortunately, most pipeline operators do not have the resources to run the complex models required to develop the numerous

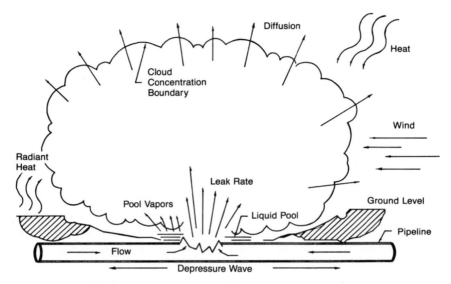

Figure 7-4. Vapor cloud from pipeline rupture.

scenarios that would be needed. So again, we turn to a few easily obtained parameters that may allow us to determine a relative risk ranking of some scenarios. An exact numerical solution is not sought.

Dispersion studies have revealed a few simplifying truths that can be used in this risk assessment. In general, the RATE of vapor generation, NOT the total VOLUME of released vapor, determines the cloud size. A cloud reaches an equilibrium state for a given set of atmospheric conditions. At this equilibrium, the amount of vapor added from the source exactly equals the amount of vapor that leaves the cloud boundary (the cloud boundary can be defined as any vapor concentration level). So when the surface area of the cloud reaches a size whereby the rate of vapor escaping the cloud equals the rate entering the cloud, the surface area will not grow any larger (Figure 7-4). The vapor escape rate at the cloud boundary is governed by atmospheric conditions. The cloud will therefore remain this size until the atmospheric conditions or the source rate change. This fact thus yields one quantifiable risk variable: leak rate.

A criterion must now be established for choosing a leak rate scenario. It is reasonable to assume that virtually any size leak may form in any pipeline. The evaluator could simply choose a 1-in. diameter hole as the leak size. This however, would not adequately distinguish between a 36-in. pipeline and a 4-in. pipeline. While a 1-in. hole in either might cause approximately the same size cloud (initially, at least), we intuitively believe that a 36-in. pipeline presents a greater hazard than does a 4-in. pipeline (all other factors being equal). This is no doubt because a much greater release can occur from the 36-in. line than from the 4-in. line.

It is suggested, therefore, that a leak scenario of a complete line failure—a guillotine-type shear failure—should be used to model the leak rate. This type of failure causes the leak rate to be calculated based upon the line diameter and pressure. Even though this type of line failure is rare, the risk assessment is still valid. By consistency of application, we can choose any hole size and leak rate. We are simply choosing one here that serves the dual role of incorporating the factors of pipe size and line pressure directly into rating vapor cloud size.

Leak rate can be approximated by calculating how much vapor will be released in ten minutes. The highest leak rate occurs when the pressure is the highest and the escape orifice is the largest. This leads to the assumption that, in most cases, the worst leak rate happens near the instant of pipeline rupture, while the internal pressure is still the highest and after the opening has reached its largest area. This highest leak produces the largest cloud. As the leak rate decreases, the cloud shrinks. In the case of a dense cloud, vapors may "slump" and collect in low lying areas or "roll" downhill as the cloud seeks its equilibrium size. We are conservatively assuming that all the vapor stays together in one cloud for the full ten-minute release. We can also conservatively, neglect the depressuring effect of ten-minutes worth of product leakage. This is done to keep the calculation simple. The ten-minute interval is chosen to allow a reasonable time for the cloud to reach maximum size, but not long enough to be counting an excessive mass of well-dispersed material as part of the cloud. The amount of product released and the cloud size will almost always be overestimated using the above assumptions. Again, for purposes of the relative risk assessment, overestimation is not a problem as long as consistency is ensured. See Appendix B for more discussion of leak rate determinations.

A second simplifying parameter is the effect of molecular weight on dispersion. Molecular weight is inversely proportional to the rate of dispersion. A higher molecular weight produces a denser cloud that has a slower dispersion rate. A denser cloud is less impacted by buoyancy effects and air turbulence (caused by temperature differences, wind, etc.) than a lighter cloud. Using this fact yields another risk variable: product molecular weight.

In the absence of more exact data, it is therefore proposed that the increased amount of risk due to a vapor cloud will be assessed based upon two key variables: leak rate and product molecular weight. Meteorological conditions, terrain, chemical properties and a host of other important variables are intentionally being omitted for two reasons. First, they are highly variable in themselves and consequently difficult to model or measure. Second, they add much complexity and, arguably, little additional accuracy, for our purposes here.

A point schedule can now be designed to quantify the increase in hazard as the dispersion characteristics of molecular weight and leak rate combine.

MW	Product Released after 10 Minutes (pounds)			
	0–5,000	5,000–50,000	50,000–500,000	>500,000
≥50	4	3	2	1
28–49	5	4	3	2
≤27	6	5	4	3

These points are the vapor spill score. In the table, the upper right corner reflects the greatest hazard, while the lower left is the lowest hazard. By the way in which the dispersion factor is used to adjust the acute or chronic hazard, a higher spill score will yield a safer condition.

By using only these two variables, several generalizations are being implied. For instance, the release of 1,000 lbs of material in ten minutes potentially creates a larger cloud than the release of 4000 lbs in an hour. Remember, it is the *rate* of release that determines cloud size, not the total volume released. The 1,000-lb release therefore poses the greater hazard than the 4,000-lb release. Also, a 1,000-lb release of MW 16 material such as methane is less of a hazard than a 1,000-lb release of MW 28 material such as ethylene. The schedule must now represent the evaluator's view of the relative risks of a slow 4,000-lb MW 28 release versus a quick 1,000-lb MW 16 release. Fortunately, this need not be a very sensitive ranking. Orders of magnitude are sufficiently close for the purposes of this assessment.

Again, the score from this table is the vapor spill score. This number will be combined with the population score and the product hazard to determine the consequences of a vapor release.

Liquid Spills

Releases of products, which for the most part remain in liquid form, pose hazards of a different nature.

While we do not attempt to quantify the range or dispersion of a liquid spill, we must recognize the increased hazard with increasing spill quantity, population proximity, and proximity of environmental receptors. A larger spill and a spill close to sensitive areas creates a greater risk of harm by ignition if the material is flammable or by direct contact with living creatures if the product is toxic.

To correctly analyze a liquid spill, a host of variables must be assessed. These include:

- product viscosity
- soil infiltration rate
- vegetative cover effects
- slope effects
- product solubility
- evapotranspiration rate
- groundwater flow patterns
- proximity to surface waters
- product miscibility

In trying to rank spill rangeabilities, it is probably not necessary to distinguish between all of these factors. The added complexities are thought to far outweigh the benefits of such detailed calculations. The problem is simplified here to two factors: maximum possible leak rate and soil permeability (or its equivalent if a release into water is being studied).

The extent of the liquid spill threat is dependent upon the extent of the spill dispersion, which in turn is dependent upon the size of the spill, the type of product spilled, and the characteristics of the spill site. The size of the spill is contingent upon the rate of release and the duration. Slow leaks gone undetected for long periods can be more damaging than massive leaks that are quickly detected and addressed. This aspect is considered indirectly by the effectiveness of pipeline patrolling and other methods of leak detection. The consequences of a slow leak appear in the chronic hazard rating of the substance.

The characteristics of the spill site help determine the movement of the product. The possibilities are spills into air, surface water, soil, and groundwater. Accurately measuring these movements is an enormously complex modelling process. For releases into the air, product movement in the form of vapor generation from the spilled liquid is covered in the discussion of vapor dispersion. Because product release from a pipeline is a temporary excursion, the pollution potential beyond immediate toxicity or flammability is not specifically addressed for releases into the air. This neglects the accumulative damage that can be done by many small releases of atmosphere-damaging substances (such as the so-called greenhouse gases that are thought to damage the ozone). Such chronic hazards are considered in the assignment of the equivalent reportable release quantity ($RQ_{equivalent}$) for volatile hydrocarbons.

Releases into surface waters are the second potential type of environmental insult. The size of the body of water and its uses determine the severity of the hazard. If the water is used for swimming, fishing, livestock watering, irrigation, or drinking water, pollution concentrations must be kept quite low. The hazards associated with spills into surface waters should be assessed on a case-by-case basis and combined with the soil permeability factor which is explained below. Because we are most concerned with the rangeability of the spill, the spill into water should take into account the miscibility of the substance with water and the water movement. A spill of immiscible material into stagnant water would be the equivalent of a relatively impermeable soil. A highly miscible material spilled into a flowing stream is the equivalent of a highly permeable soil.

Spills in soil or rock are the most common pipeline environmental concern. Such spills also carry the potential for groundwater contamination.

Product movement through the soil is dependent upon such soil factors as adsorption, percolation, moisture content, and bacterial content. Soil characteristics can be best assessed by using one of the common soil classification systems, such as the USDA soil classification system which incorporates physical, chemical, and biological properties of the soil. For simplicity, only one soil characteristic is considered in this risk evaluation. This is also the soil characteristic that is used in the EPA Hazard Ranking System (HRS)—permeability of geologic materials [4].

The following table can be used to score the soil permeability for liquid spills into the soil:

Table 7-3
Soil Permeability Score

Description	Permeability	Point Score
impervious barrier	0 cm/sec	5 points
clay, compact till, unfractured rock	$< 10^{-7}$ cm/sec	4 points
silt, silty clay, loess, clay loams, sandstone	10^{-5} to 10^{-7} cm/sec	3 points
fine sand, silty sand, moderately fractured rock	10^{-3} to 10^{-5} cm/sec	2 points
gravel, sand, highly fractured rock	$> 10^{-3}$ cm/sec	1 point

This soil permeability factor will be one part of the liquid spill score. Together with a ranking of the spill size, the liquid spill score will be determined. Ultimately, a scoring of the spilled substance's hazards and persistence (considering biodegradation, hydrolysis, and photolysis), will combine with this number in evaluating the consequences of the spill.

We now need to distinguish the size of the spill. We assume that larger spills are more hazardous than smaller spills. Spill size is a variable dependent upon the system hydraulics and the reliability and reaction times of safety equipment and pipeline operators. Safety equipment and operation protocol are covered in other sections of the assessment, so the system hydraulics alone will be used here to rank spill size. We will include an adjustment to the spill score when it can be shown that special facilities exist which will reliably reduce the potential spill size by at least 50%.

Leak rate is determined with a worst case line break scenario. As with the atmospheric dispersion, choosing this scenario allows us to incorporate the line size and pressure into the hazard evaluation. A 36-in. high pressure gasoline line poses a greater threat than a 4-in. high pressure gasoline line, all other factors being equal. This is because the larger line can potentially create the larger spill. The leak rate should include product flow from pumping equipment. Reliability of pump shutdown following a pipeline failure is considered elsewhere.

Because the release of a small amount of an incompressible liquid will depressure the pipeline quickly, the longer-term driving force to feed the leak may be gravity and siphoning effects. A leak in a low-lying area may be fed for some time by the draining of the rest of the pipeline. The evaluator should find the worst case leak location for the section being assessed.

Based upon the worst case leak rate and leak location for the section, the spill size can be ranked according to how much product is spilled in a period of one hour. The one-hour period is somewhat arbitrary, but will serve our purposes for a relative ranking. Leaks can be (and have been) allowed to continue for more than one hour. Leaks can also be quickly isolated and contained. This approach will, however, distinguish the more hazardous situations such as high pressure, large diameter pipelines in low-lying areas.

Points can be assessed based upon the quantity of product spilled, under a worst case scenario, in one hour:

Pounds Spilled	**Point Score**
< 100	5 points
101–1000	4
1001–10,000	3
10,001–100,000	2
100,000	1

The spill size points are then averaged with the soil permeability points to arrive at the liquid spill score. This number will then be adjusted if the leak detection and emergency response activities can ensure a 50% reduction in spill size or dispersion.

Liquid Spill = [(spill size) + (soil permeability)] ÷ 2
                    + (adjustment factor)

Example:

A 12-in. crude oil pipeline is being evaluated. In this section, the line is in a valley of sandy soil. It crosses beneath a river at its lowest elevation.

The evaluator assesses the situation as follows: A complete line rupture at its lowest point (beneath the river) would allow the maximum spill volume. With continuous pumping and the effects of gravity, this spill would be well over one million pounds in an hour. The spill onto the sandy soil would score 1 point on the soil permeability scale. The spill into the river would also score the maximum of 1 point because the river flow would ensure wide dispersion of the product, even though it is fairly immiscible in water. Either way, the liquid spill would be scored as 1 point. The worst case is then scored as:

$$\frac{\begin{array}{r}\text{Spill size} = 1 \text{ point}\\ \text{Soil permeability} = 1 \text{ point}\end{array}}{\text{Average} = 1 \text{ point}}$$

Note: Methods to adjust this score based upon emergency response and leak detection activities are discussed in the next section.

Adjustments to Liquid Spill Score

As previously defined, the chronic hazards have a time factor implied—as time goes on, the hazard becomes worse. Actions that can influence this time factor will therefore impact the chronic hazard. These actions must do one of three things:

- limit the amount of spilled product
- limit the area of opportunity for consequences
- limit the loss or damage caused by the spill

Limiting the amount of product spilled is done by isolating the pipeline quickly. The area of opportunity is limited by evacuating people and animals, by removing possible ignition sources, or by containing the spill. Loss is limited by prompt medical attention, quick containment, and cleanup of the spill.

Two pipeline activities that can contribute to chronic hazard reduction are Leak Detection and Emergency Response. The amount of the contribution to the overall risk picture is arguable. One of the fastest detection and response scenarios would be valves that automatically isolate a leaking pipeline section. By one study of 336 liquid pipeline accidents, such valves could, at best, have provided a 37% reduction in damage [25]. The authors calculate that the costs (installation and ongoing maintenance) would far outweigh the possible benefits, and also imply that such valves may actually introduce new hazards [25].

By the proposed method of spill size determination, an 8-in. pipeline presents a greater hazard than does a 6-in. pipeline (all other factors held constant). When the leak detection/emergency response actions can limit the spill size from an 8-in. line to the maximum spill size from a 6-in. line, some measure of risk reduction has occurred.

There is little argument that, *under the right conditions,* leak detection and emergency response can indeed reduce risk. They are therefore included as modifiers to the liquid dispersion portion of the leak impact factor. This is an all or nothing adjustment to the liquid spill score, based on the actions' potential ability to achieve a 50% reduction in the consequences of the risk.

Leak Detection

Leak detection can be seen as a part of emergency response. It merely provides early notification of a hazardous event, and hence allows more rapid response to that event. Leak detection is however a reaction to an event—the leak—that has already occurred.

Pipeline leak detection can take a variety of forms, several of which have been previously discussed. The most common is direct observation. Leak sightings by pipeline employees, neighbors, and the general public as well as sightings while patrolling or surveying the pipeline are examples of direct observation leak detection. Pipeline patrolling or surveying can be made more sensitive by adjusting observer training, speed of survey or patrol, equipment carried (may include gas detectors, infrared sensors, etc.), altitude/speed of air patrol, topography, ROW conditions, product characteristics, etc. While these are sometimes inexact, experience shows them to be rather consistent leak detection methods.

Other forms of leak detection include 1) pressure sensing, where abnormally low pressures or abnormal rate-of-change of pressure are detected, and 2) flow rate analysis, where flow rates into a pipeline section are compared with flow rates out of the section and discrepancies are detected. Both of these methods may have sensitivity problems because they must not give leak indications in cases where normal pipeline transients (unsteady flows or pressures, sometimes temporary as the system stabilizes after some change is made) are causing pressure swings and flow rate changes. Generally the operator must decide between many false alarms and low sensitivity to actual leaks. Because pipeline leaks are hopefully rare-occurrence events, the latter is often chosen.

More sophisticated leak detection methods require more instrumentation and computer analysis. One method is designed to detect pressure waves. A leak will cause a negative pressure wave at the leak site. This wave will travel in both directions from the leak at high speed through the pipeline product (much faster in liquids than in gases). By simply detecting this wave, leak size and location can be estimated. A technique called Pressure Point Analysis (PPA) detects this wave and also statistically analyzes all changes at a single pressure or flow monitoring point. By statistically analyzing all this data, the technique can reportedly, with a higher degree of confidence, distinguish between leaks and many normal transients as well as identify instrument drift and reading errors.

Another method is a computer-based technique that uses SCADA data in conjunction with mathematical algorithms to analyze pipeline flows and pressures on a real-time basis. Conservation of momentum calculations, conservation of energy calculations, and a host of sophisticated flow equations are generally used. The more instruments that are accurately transmitting data into the model, the higher the accuracy of the model and the confidence level of leak indications. Ideally, the model would receive data on flows, temperatures, pressures, densities, viscosities, etc. along the entire pipeline length. By tuning the computer model to simulate mathematically all flowing conditions along the entire pipeline and then comparing this simulation to actual data, the model tries to distinguish between instrument errors, normal transients, and leaks. Reportedly, small leaks can be accurately located in a timely fashion. How small a leak and how swift a detection is specific to the situation.

A final method of leak detection involves the installation of a secondary conduit along the entire pipeline length. This secondary conduit is designed to sense leaks originating from the pipeline. The secondary conduit may take the form of a small diameter perforated tube, installed parallel to the pipeline, which allows air samples to be drawn into a sensor which can detect the product leaks. The conduit may also totally enclose the product pipeline and allow the annular space to be tested for leaks. Obviously these systems can cause a host of logistical problems and are usually not employed except on short lines.

The method of leak detection chosen is dependent upon a variety of factors including the type of product, flow rates, pressures, the amount of instrumentation available, the instrumentation characteristics, the communications network, the topography, the soil type, and economics. As previously mentioned, when highly sophisticated instruments are required, there is often a tradeoff between the sensitivity and the number of false alarms, especially in noisy systems with high levels of transients.

At this time, instrumentation and methodology designed to detect pipeline leaks impacts only a narrow range of the risk picture. Detection of a leak obviously occurs after the leak has occurred. As is the case with other aspects of emergency response, leak detection is thought to normally play a minor role, if any, in reducing the hazard, reducing the probability of the hazard, or reducing the acute consequences. Leak detection can play a larger role in reducing the chronic consequences of a release.

One can imagine a scenario in which a small leak, rapidly detected and corrected, averted the creation of a larger, more dangerous leak. This would theoretically reduce the acute consequences of the leak. We can also imagine the case where rapid leak detection coupled with the fortunate happenstance of pipeline personnel being close by, might cause reaction time to be swift enough to reduce the extent of the hazard. This would also impact the acute consequence factor. These scenarios are obviously rare. Increasing use of leak detection methodology is to be expected as techniques become more refined and instrumentation becomes more accurate. As this happens, leak detection may play an increasingly important role in risk management.

The evaluator should assess the nature of leak detection abilities in the pipeline section he is evaluating. The assessment should include:

- what size leak can be reliably detected
- how long before positive detection as a leak
- how accurately can the leak location be determined

Larger leaks can be detected more quickly and located more precisely. Smaller leaks may not be found at all by some methods due to the sensitivity adjustments. The tradeoffs involved between sensitivity and leak size are usually expressed in terms of uncertainty.

Emergency Response

As is the case in leak detection, there is a category of scenarios where an improved emergency response can reduce a risk. Because this response is also a reaction to an event that has already occurred, its impact will normally occur only in the consequences portion of the risk.

The most probable pipeline leak scenarios suggest that the leak rate would not grow larger over time because the driving force (pressure) is being reduced immediately after the leak event begins. (The exception is the case where a liquid pipeline may have an elevation profile that supports a sustained leak rate.) This means that reaction times swift enough to impact the immediate degree of hazard are not very likely. We emphasize immediate here so as not to downplay the importance of emergency response. Emergency response can indeed influence the final outcome of an event in terms of loss of life, injuries, and property damage. This is not thought to impact the acute hazard, however. A spill with chronic characteristics, where the nature of the hazard causes it to increase in severity as time passes, can be impacted by emergency response. In these cases, emergency response actions such as evacuation, blockades, and rapid pipeline shut-off are effective in reducing the hazard.

The evaluator should examine the response possibilities and the most probable response scenario. If he determines that the emergency response actions will reliably reduce the leak consequences by 50%, he should adjust the chronic leak impact factor by one point.

Actions that may accomplish this are:

Spill Limiting Actions

This is the most realistic way for the operator to be able to reduce the spill impact by 50%. The evaluator should refer to the Safety Systems item in the *Incorrect Operations Index* to help verify the effectiveness of mechanical spill-limiting devices.

A. Automatic valves. Set to close automatically, these valves are often triggered on low pressure, high pressure, high flow, or rate of change of pressure or flow. Regular maintenance is required to ensure proper operation. Experience warns that this type of equipment is often plagued by false trips that are sometimes cured by setting relatively insensitive response trigger points.

   Check valves are another form of automatic valves and play a spill reducing role. A check valve might be especially useful for liquid lines with elevation changes. Strategically placed check valves may reduce the draining or siphoning to a spill at a lower elevation.

   Included in this section should be automatic shutoffs of pumps, wells, and other pressure sources. Redundancy should be included in all such systems before risk-reducing credit is awarded (see *Incorrect Operations Index*).

B. Valve spacing. Close valve spacing may provide a benefit in reducing the spill amount. This must be coupled with the most probable reaction time in closing those valves.

C. Sensing devices. Part of the equation in response time is the first opportunity to take action. This opportunity is dependent upon the sensitivity of the leak detection. All leak detection will have an element of uncertainty, from the possibility of crank phone calls to the false alarms generated by instrumentation failures or instrument reactions to pipeline transients. This uncertainty must also be included in the following item.

D. Reaction times. If an operator intervention is required to initiate the proper response, this intervention must be assessed in terms of timeliness and appropriateness. A control room operator must often diagnose the leak based upon instrument readings transmitted to him. How quickly he can make this diagnosis is dependent upon his training, his experience, and the level of instrumentation that is supporting his diagnosis. Probable reaction times can be judged from mock emergency drill records when available. The evaluator can incorporate his *Incorrect Operations Index* ratings (training, SCADA, etc.) into this section also.

If the control room can remotely operate equipment to reduce the leak, the reaction time is obviously improved. Travel time by first responders must otherwise be factored in. If the pipeline operator has provided enough training and communications to public emergency response personnel so that they may operate pipeline equipment, response time may be improved, but possibly at the expense of increased human error potential. Public emergency response personnel are probably not able to devote much training time to a rare event such as a pipeline failure.

If the reaction is automatic (computer generated valve closure, for instance) a sensitivity is necessarily built in to eliminate false alarms. The time it takes before the shut down device is certain of a leak must be considered.

"Area of Opportunity" Limiting Actions

A. Evacuation. Under the right conditions, emergency response personnel may be able to safely evacuate people from the spill area. In order to do this, they must be trained in pipeline emergencies. This includes having pipeline maps, knowledge of the product characteristics, communications equipment, and the proper equipment to enter into the danger area (breathing apparatus, fire retardant clothing, hazardous material clothing, etc.). Obviously, entering a dangerous area in an attempt to evacuate people is a situation-specific action. The evaluator should look for evidence that emergency responders are properly trained and equipped to exercise any reasonable options after the situation has been assessed. Again the criteria must include the time factor. Credit is given when the risk can be reliably reduced by 50% due to appropriate emergency response actions.

B. Blockades. Another limiting action in this category is to limit the possible ignition sources. Preventing vehicles from entering into the danger zone has the double benefit of reducing human exposure and reducing ignition potential. The blockade must be in place in time to reduce the risk by 50%.

C. Containment. Especially in the case of restricting the movement of hazardous materials into the groundwater, quick containment can reduce the consequences of the spill. The evaluator should look for evidence that the response team can indeed reduce the spreading potential by 50%. This is usually in the form of secondary containment.

Loss Limiting Actions

Medical Treatment. Proper care of persons affected by the spilled product may reduce losses. Again, product knowledge, proper equipment, proper training, and quick action on the part of the responders are necessary factors.

Other items that play a role in achieving the consequence-limiting benefits include the following:

- emergency drills
- emergency plans
- communications equipment
- proper maintenance of emergency equipment
- updated phone numbers readily available
- extensive training including product characteristics
- regular contacts and training information provided to fire departments, police, sheriff, highway patrol, hospitals, emergency response teams, government officials

These can be thought of as characteristics that help to increase the chances of correct and timely responses to pipeline leaks. Perhaps the first item, mock drills, is the single most important characteristic. It requires the use of many other list items and demonstrates the overall degree of preparedness of the response efforts.

Many factors have been mentioned here. By necessity, the evaluation of effectiveness will be situation-specific. At first look, it may appear that an operator has many of these systems in place and functioning to a high level. Realistically, however, it is difficult to meet the criteria of a 50% reduction in the effective spill size. The spill size is calculated as the amount of product spilled in one hour, assuming worst cases. To reduce this, actions would have to ALWAYS take place long before this one-hour time period.

The evaluator can take the following approach to tie this together. An example follows.

Step 1:   The evaluator uses the worst case pipeline spill scenario or a combination of scenarios to work from. He calculates the worst case one-hour liquid spill size.

Step 2:   The evaluator determines, with operator input, methods to attain a 50% risk reduction such as reduce spill amount by 50%, reduce population exposure by 50% (number of people or duration of exposure), contain 50% of spill before it can cause damage, reduce health impact by 50%.

Step 3:   The evaluator determines if any action or combination of actions can reliably reduce the risk by 50%. This is done with consideration given to the degree of response preparedness.

If he decides that the answer in Step 3 is *yes,* he then adds one point to the Liquid Spill Score ([spill size + soil permeability] ÷ 2) calculated earlier.

It may appear that much time and energy is being spent on the determination of an adjustment of only one point. Remember that the leak impact factor is a multiplier, so one point can have a significant effect on the overall risk score.

Example A:

The evaluator is assessing a section of gasoline pipeline through the town of Smithville.

The scenario he is using involves a leak of the full pipeline flow rate. This hypothetical leak occurs at a low point in the line profile, in the center of Smithville. He recognizes the acute hazard of flammability and the chronic hazards of toxicity (high benzene component), residual flammability (from pockets of liquid), and environmental insult. He feels that a 50% reduction in risk can be attained if the spill is reduced by 50%, if 50% of the spilled product is contained quickly, or if 50% of the potentially affected residents can be evacuated BEFORE they are exposed to the acute hazard.

He has determined that the leak detection and emergency response activities are in place to warrant an adjustment of the chronic leak impact factor (one point is added).

The basis for this determination is the following items observed or ascertained from interviews with the operators:

- automatic valves are set to isolate pipeline sections around the town of Smithville. The valves trigger upon a pressure drop of more than 20% from normal operating conditions. The valves are thoroughly tested every six months and have a good operating history. A 20% drop in pressure would occur very soon after a substantial leak.

- annual emergency drills are held, involving all emergency response personnel from Smithville. The drills are well-documented and reflect a high degree of response preparedness.

Presence of the automatic valves should limit the spill to 50% of what it would be without the valves. This alone would have been sufficient to adjust the chronic leak impact factor. The strong emergency response program should limit exposure due to residual flammability and ensure proper handling of the gasoline during cleanup. Containment is not seen as an option, but by limiting the spill size, the environmental insult is minimized also. The evaluator sees no relief from the acute hazard, but feels an adjustment for the chronic hazard is appropriate.

Example B:

The evaluator is assessing a section of brine pipeline in a remote, unpopulated area.

The leak scenario he is using involves a complete line rupture. The hazards are only chronic in nature—no immediate threats to public or responders. The chronic threat is the exposure to the groundwater table which is shallow in this area.

The best chance to reduce the chronic risk by 50% is seen as limiting the spill size by 50%. Emergency response will not reliably occur quickly enough to isolate the leaking pipeline before line depressurization and pump shutoffs slow the leak anyway. Containment in a timely fashion is not possible.

No adjustments to the chronic leak impact factor are made.

The vapor spill score or the liquid spill score will now be used to continue the calculation of the Dispersion Factor.

Population Density

As part of the consequence analysis, a most critical parameter is the proximity of people to the pipeline failure. This impacts both the acute and the chronic hazards. The impact on the flammability and toxicity hazards is self-evident. The impact on the chronic hazards may be more subtle. Population proximity is a factor here because the area of opportunity for harm is increased as human activity is closer to the leak site. Potential for ingesting contaminants through drinking water, vegetation, fish, or other ingestion pathways is higher when the leak site is nearby. Less dilution has occurred and there is less opportunity for detection and remediation before the normal pathways are contaminated. The other pathways, inhalation and dermal contact, are similarly impacted.

Population density is taken into account by using the DOT Part 192 class locations, 1, 2, 3, and 4. These are for rural to urban areas, respectively. The class locations are determined by examining the area 660 ft on either side of the pipeline centerline, and one mile along the pipeline. This one mile by 1,320 ft rectangle, centered over the pipeline, is the defined area. For purposes of the class definitions, it should be thought of as an area continuously moving along the pipeline.

If any one-mile stretch of pipeline has more than 46 dwellings inside this defined area, that section is termed to be in a Class 3 Area. A section with less than 46 dwellings but more than 10 dwellings in the defined area is termed to be in a Class 2 Area. Each mile with less than 10 dwellings is considered to be in a Class 1 Area. A Class 4 Area exists when the defined area has a prevalence of multi-story buildings.

A Class 3 Area is also defined as a section of pipeline that has a high-occupancy building or well-defined outside meeting area within the defined area. Buildings such as churches, schools, and shopping centers that are regularly occupied (5 days per week or 10 weeks per year) by 20 or more people are deemed to be high-occupancy buildings. The presence of one of these within 660 ft of the pipeline is sufficient condition to classify the pipeline section as Class 3. Note that this building will effectively cause two miles of pipeline to be classified as Class 3 because it will be located in one-mile sections in both directions along the pipeline.

Points are awarded based upon the class location:

Class 1 .... 1 point
Class 2 .... 2 points
Class 3 .... 3 points
Class 4 .... 4 points

These points will be an adjustment to the spill score which will change the *Dispersion Factor* and hence the *Leak Impact Factor*.

Dispersion Factor = (Spill Score)/(Population Score)
$\qquad$ = 1/4 = 0.25 Worst case
$\qquad$ = 6/1 = 6.0 Best case
Leak Impact Factor = (Product Hazard)/(Dispersion Factor)
$\qquad$ = 22/0.25 = 88 Worst case
$\qquad$ = 1/6 = 0.2 Best case

By the above relationships, we see that the dispersion factor can change the acute and chronic hazards enormously. This in turn changes the *Leak Impact Factor* enormously, which will ultimately determine the relative risk score.

CHAPTER EIGHT

Managing the Data

Data Display and Manipulation

Computerize It!

The computer can become a most valuable tool in pipeline risk assessment. Because a great deal of information can be gathered for each pipeline section evaluated, it does not take many evaluations before the total amount of data becomes unwieldy. The computer is a useful way to store and, more importantly, retrieve and organize the data. The potential for errors in number handling is reduced if the computer performs the calculations to arrive at the index values, the *Leak Impact Factor,* and the final risk value.

Almost any programming language could be used to handle the data input and calculations. As the database grows, the need for programs or routines that can quickly and easily (from the user standpoint) search a database and display the results of the search becomes more important. Flexibility in database searches and displays is very useful. For all these reasons, a spreadsheet program may be one of the best choices for this application. Spreadsheet software is readily available, widely understood, and well-suited to data entry and display. Most spreadsheets have database routines. Where the capabilities of these routines must be enhanced, additional database routines or separate database programs can usually be attached to the spreadsheet program.

Before the computer program is created, the programmer should do some design work. He must have a good understanding of how the program is going to be used and by whom—the software should be designed with the customer in mind. Programs often get used in ways slightly different from the original intentions. The most powerful software has successfully anticipated the user's needs, even if the user himself has not anticipated every need! Data input and the associated calculations are rather straightforward. Database searches, comparisons, and displays are highly use-specific. The programmer will do well to invest some time planning and anticipating user needs.

There will most likely be several ways in which the data needs to be sorted and displayed. This is again dependent upon the intended use. Some potential applications are discussed below.

Application 1: Risk Awareness

This is most likely the driving force behind performing risk evaluations on a pipeline system. Owners and/or operators want to know how their systems rate from a risk standpoint. This rating is perhaps best presented in the form of a rank-ordered list. The rating or ranking list should include some sort of reference point—a baseline or standard to be used for comparisons. The reference point, or standard, gives a sense of scale to the rank ordering of the company's pipeline sections.

The standards may be based upon:

(1) governing regulations, either from local government agencies or from company policies

(2) a pipeline section or sections that are intuitively thought to be safer than the other sections

(3) a fictitious pipeline section—perhaps a low pressure nitrogen or water pipeline in an uninhabited area for a low risk score, perhaps a high pressure hydrogen cyanide (very flammable and toxic) pipeline through a large metropolitan area for a high risk score.

By including a standard, the user sees not only a rank ordered list of his facilities, he also sees how the whole list compares to a reference point that he can understand.

Ideally, the program to support Application 1 will run something like this:

Data is input for the standard and for each section evaluated. The computer program calculates numerical values for each index, the Leak

Impact Factor (product hazards and spill scores), and the final risk rating for every section. Any of these calculations may later be required for detailed comparisons to standards or to other sections evaluated. Consequently, all data and intermediate calculations must be preserved and available to search routines. The program will likely be called upon to produce displays of pipeline sections in rank order. Sections may be grouped by product handled, by geographic area, by index, by risk rating, etc.

Examples

There are countless ways in which the risk picture may need to be presented. Following are three examples of Application 1 needs.

1) Pipeline company management wants to see the 20 most hazardous sections operated by the company. A list is generated, ranking all sections by their final relative risk number. A bar chart provides a graphic display of the 20 sections and their relative magnitude to each other.

2) Pipeline company management wants to see the 20 most hazardous sections in natural gas service in the state of Oklahoma. A rank-ordered list for natural gas lines in Oklahoma is generated.

3) The corrosion control department wants to see a rank ordering of all sections, ranked by corrosion indexes, lowest to highest. All pipeline sections are ranked strictly by Corrosion Index score.

Application 2: Compliance

Another anticipated application of this program is a comparison to determine compliance with local regulations or with company policy. In this case, a standard is developed based upon the company's interpretation of government regulations and upon the company policy for the operation of pipelines (if that differs from regulatory requirements). The computer program will most likely be called upon to search the database for instances of non-compliance with the standard(s).

To highlight these instances of non-compliance, the program must be able to make correct comparisons between standards and sections evaluated. Liquid lines must be compared with liquid regulations; Texas pipelines must be compared with Texas regulations, etc.

If the governing policies are performance-based (corrosion must be prevented . . . all design loadings anticipated and allowed for . . . etc.), the standard may change with differing pipeline environments. It is a useful technique to pre-define the pipeline company's interpretations of regulatory requirements and company policy. These definitions will be the prevention items in the risk evaluation. They can be used to have the computer program automatically create standards for each section evaluated.

Using the distinction between attributes and preventions, a floating standard can be developed. In the floating standard, the standard changes with changing attributes. The program is designed so that a pipeline section's attributes are identified and then preventions are assigned to those attributes based upon company policies. The computer can thus generate standards based upon the attributes of the section and the level of preventions required according to company interpretations. The standard changes, or floats, with changes in attributes or company policy.

Example:

A company has decided that an appropriate level of public education is to be mailouts, advertisements, and speaking engagements for urban areas, and mailouts with annual landowner/tenant visits for rural areas. With this definition, the computer program can now assign a level of preventions of 7 points for the urban areas and a level of 6 points for rural areas. The program generates these standards by simply identifying the population density value and assigning the points.

By having the appropriate level of preventions pre-assigned into the computer, consistency is ensured. When policy is changed, the standards can be easily updated. All comparisons between actual pipeline sections and standards will be instantly updated and, hence, based on the most current company policy.

It is reasonable to assume that whenever an instance of non-compliance is found, a detailed explanation will be required. The program can be designed to retrieve the whole record and highlight the specific item(s) that caused the non-compliance.

As policies and regulations change, it will be necessary to change the standards. Routines that allow easy changes will be useful.

Application 3: What-if Trials

A useful feature in the computer program will undoubtedly be the ability to perform *what-if* trials. Here, the user can change items within each index to see the effect on the risk picture. For example, if air patrol frequency is increased, how much risk reduction is obtained? What if an internal inspection device is run in this section? If we change our public education program to include door-to-door visits, how does that influence the risk of third party damage?

It will be important to preserve the original data during the what-if trial. The trial will most likely need to be done outside the current database. A secondary database of proposed actions and the resulting risk ratings could be built and saved using the what-if trials. This second database might be seen as a target or goal database, and could be used for planning purposes.

The program should allow specific records to be retrieved as well as general groups of records. The whole record or group of records will need to be easily modified while preserving the original data. Comparisons or before and after studies will probably be desirable. Graphic displays will enhance these comparisons.

Application 4: Spending Prioritization

As an offshoot to the ranking list for relative risk assessment, it will most likely be desirable to create rank-order lists for prioritizing spending on pipeline maintenance and upgrades. The list of lowest scored sections from a corrosion risk standpoint should receive the largest share of the corrosion control budget, for instance. The spending priority lists will most likely be driven by the rank-ordered relative risk lists, but there may be the need for some flexibility. Spending priority lists for only natural gas pipelines may be needed, for example. The program could allow for the rearrangement of records to facilitate this.

A special column, or field in the database, may be added to tabulate the projected and actual costs associated with each upgrade. Costs associated with a certain level of maintenance (prevention) activities could also be placed into this field.

The user may want to analyze spending for projects on specific pipeline sections. He may alternatively wish to perform cost-benefit analyses on the effects of certain programs across the whole pipeline system. For instance, if the *Third Party Damage Index* is to be improved, the user may study the effects of increasing the patrol frequency across the whole system. The costs of the increased patrol could be weighed against the aggregate risk reduction, perhaps expressed as a percentage reduction in the sum or the average of all the risk values. This could then be judged against the effects of spending the same amount of money on, say, close interval surveys or operator training programs.

The cost-benefit studies are not absolute because this risk assessment program yields only relative answers. For a given pipeline system, however, relative answers are usually the most meaningful. The program should help the user decide where his dollar spent has the greatest impact on risk reduction.

Application 5: Comparisons

In some of the above applications and as a stand-alone application, comparisons between records will be useful. A user may wish to make a detailed comparison between a standard and a specific record. He may wish to see all items that exceed the standard or all items that are less than their corresponding standard value.

Groups of records may also need to be compared. For example, all Texas pipelines could be compared with all Louisiana pipelines or *Corrosion Index* values of natural gas pipelines could be compared with *Corrosion Index* values of crude oil pipelines. Graphics would again enhance the presentation of the comparisons.

Properties of the Software Program

The risk assessment outlined in this book is a dynamic tool. It must be kept current. It can play a significant role in all planning and decision making. The degree of use of this tool is directly related to the user-friendliness of the software that supports it.

The programmer should go to great lengths to make data entry and output generation simple. Keystrokes should be reduced, use of menus increased, and redundant operations eliminated.

Changes. Because the tool is designed to be dynamic—changing with changing conditions and new information—the software program must easily facilitate these changes. New regulations will require corresponding changes to the standards. Maintenance and upgrade activities will most likely require index items to be changed. Changes in operating philosophies or the use of new techniques will affect index items. New pipeline construction will require that new records be built. Increases in population densities will affect the *Leak Impact Factor.* The relative weighting of index items might also be subject to change.

The ability to quickly and easily make changes may be the most critical characteristic to be programmed into the tool. As soon as updates are no longer being made, the tool loses its usefulness.

For instance, suppose new data is received concerning the condition of coating for several pipeline sections. The user should be able to input the data in one place and easily mark all records that are to be adjusted with the new information. He should be able to change only the "coating condition" item without affecting any other data in any record. With only one or two keystrokes, the marked records should be updated and recalculated. The date and author of the change should be noted somewhere in the program for documentation purposes.

Searches. In most applications, it will be necessary to find specific records or groups of records. Most database routines make this easy. Normally the user specifies the characteristics of the record or records he is seeking. These characteristics are the search parameters the computer will use to find the record(s) of interest. User choices are made within fields or categories of the data. For instance, some fields that will be frequently used in database searches include:

- product type
- geographical area
- line size
- *Leak Impact Factor*
- index values

When the user performs searches, he chooses specifics within each field; natural gas/South Texas/4 in. through 12 in./all *Leak Impact Factors/Corrosion Indexes* with point values <50. It is important to show what the possible choices are in each field.

The choices must usually be exact matches with the database entries. Menus are useful here.

The user may also wish to do specific searches for a single item within an index such as find pipe factors > 1.6 or find public education programs $= 15$ pts. It is useful if the user can specify ranges when he is searching for numerical values, for example, hydrotest values from 5 to 15 points or hydrotest values < 20 pts.

Ideally, the user would be able to perform searches by defining search parameters in general fields, but still have the option of defining specific items. It would be cumbersome to prompt the user to specify each item in every field prior to a search. He should be able to quickly bypass fields in which he is not interested. An acceptable solution would be to have more than one level of fields. An upper, general level would prompt the user to choose one or more search parameters, perhaps from the example list above. He may also then choose the next level of fields if he wishes to specify more detailed parameters.

Tracking. It may be desirable to design the program to automatically track certain items. Overall changes in the risk picture, changes in indexes, or changes in the scoring of specific items may be of interest. Tracking of such changes over time shows progress towards goals.

Graphics. A picture is worth a thousand words. More importantly, though, pictures reveal things about the data that may otherwise go unnoticed. Bar graphs, pie charts, and run charts illustrate and compare the data in different ways. These graphics are easy to generate on most spreadsheet programs. Routines should be built to automatically produce the pictures—perhaps from menu selections.

Comparisons. The program should be able to display two or more records for comparison purposes. The program may be designed to highlight differences between records of certain magnitudes, for instance, highlight an item when it differs by more than 10% from the corresponding standard value.

Comparisons between groups of records may require the program to calculate averages, sums, or standard deviations.

Records being compared will need to be accessible to the graphics routines, since the graph is often the most powerful method of illustrating the comparisons.

Documentation. This book may provide most of the documentation necessary for the software program. It contains explanations as to why and how certain items are given more points than others and why certain items are considered at all. Of course, if the risk assessment used deviates from the book, explanations for all items should be provided. Supplemental documentation may be required to explain the calculation routines such as *Pipe Safety Factor* and *Leak Impact Factor*. While the book explains the reasoning and provides the necessary equations, the programmer should document the location of each equation and the location from which any variables are obtained.

Protection. The database should be protected from tampering. Access to the data can generally be given to all potential users, while withholding change privileges. Because all users will be encouraged to understand and use the program, they must be allowed to manipulate data, but this should probably be done exclusive of the main database. An individual or department can be responsible for the main database. Changes to this main database should only be made by authorized personnel, perhaps through some type of formal change-order system.

Building the Program

Once the programmer understands how the program is going to be used and which options will be of most use to users, he can start programming. Suggestions are offered for the programmer who is creating the program in a spreadsheet environment.

Figure 8-1 shows a typical layout of the program. The spreadsheet regions are offset to allow for insertion of rows and columns without affecting other regions.

Map and Macros

This first region contains a map of the entire spreadsheet. This map should serve as a directory to show where the various components of the program are located. The map may be a list or a graphic representation similar to Figure 8-1. It should provide spreadsheet addresses for quick access to the other regions. While movements within the spreadsheet will probably be controlled (via menus or

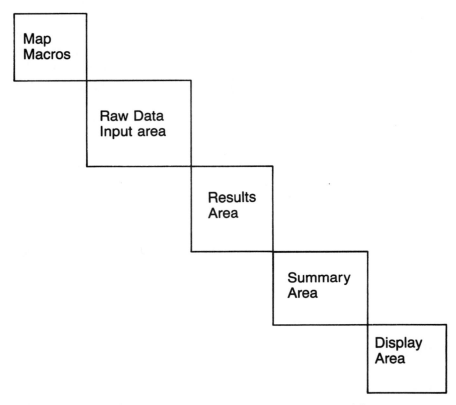

Figure 8-1. Example layout of computer spreadsheet.

prompts) for the casual user, this map will help programmers to troubleshoot and to make changes in the programs.

The first region also contains a list of all spreadsheet routines. Spreadsheet routines can be used to control movements through the program. This includes the use of menus. These routines may also perform calculations and move data between regions. Searches will be done with these routines—search parameters are prompted, search is performed, results are displayed.

By keeping all the routines together, troubleshooting is made easier. Changes in the program can also be made in one region this way. Enough documentation should accompany each routine to thoroughly describe its function.

Some examples of routines that may be used are listed in Figure 8-2.

Raw Data Input

This area of the spreadsheet should look similar to the overall view of the program shown in the Introduction. A template is built here so that all data for a given section can be inputted. This is the only area of the spreadsheet where data is entered. All calculations obtain variables from this region. After input, the data will be appropriately transferred to another region of the spreadsheet where more calculations are performed and a final record for the pipeline section is built.

When a record is to be changed, the record is brought into this region, changes are made, and then the revised record is transferred back into the database.

Often several pipeline sections will be essentially the same except for a few items. Routines should be developed to allow for easy duplication of records. That way, the few items which are different can be changed without the need to rebuild the entire record.

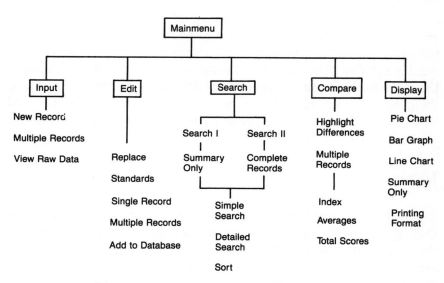

Figure 8-2. Examples of top level menus.

Results

The results are stored in the working database. It may reside outside the spreadsheet if it is too large. The database should be accessible from the spreadsheet or at least linked in some fashion. This facilitates the movement of records to and from the calculation regions.

Records can be arranged in columns or in rows, depending upon how the database reads the data. If arranged in rows, the columns become the fields; alternately, the rows become fields for records arranged in columns. Field titles and record descriptions should be included. For changes to many records at once, the change can often be done in the database itself, without having to retrieve records into the data input region.

Summary

A smaller database can be set up in conjunction with the Results region. This smaller database would contain summary information for each record. The summary may include a description of the record, the four index values, the *Leak Impact Factor,* and the final relative risk rating.

In many cases, this type of summary information will be all that is required by a user. The option should be included, however, to link the summary of the record with the complete record so the user can quickly retrieve more detailed information if desired. Depending upon the database program, a separate database may not be necessary. If the program is set up to do so, summary data only can be retrieved by designing the program to retrieve only those fields that contain the summary data.

Display

Finally, a region is set up to display the results of database searches. One of the spreadsheet movement routines should bring this region to the screen after the search routine is completed. A template, or title row, may be useful if field names are not brought into this region with the records.

Menus

As mentioned throughout this discussion, menus are often the best vehicle to allow the user quick access to the program routines. Not only do they make the program more user-friendly, they also protect the routines by controlling user movements.

Below is a list of some of the menus which may be appropriate (Figure 8-2).

Mainmenu. This menu can serve as a starting point for the user. It should contain a listing of all the options available to the user. It is the top level menu and provides an access to lower level menus.

Input. The user selects this menu to begin data entry. From here, he calls up the template region of the spreadsheet where the data on each section is to be inputted. He should be able to recall a previous record into the template region if a minor change is required. This recall option may require a linkage to the search routines.

Search. This menu prompts the user to specify the parameters for the fields that drive the search. Depending upon the database search program, this menu might offer choices of two or more search routines depending upon the level of detail of the search.

Edit. As previously mentioned, the ability to make nearly effortless changes to the database is critical to the success of the program. This menu allows access to routines that can change the entire database based upon a change in one or more fields (perhaps due to a change in company policy or governing regulation). It should also provide routines to make minor changes to individual records or groups of records. Again, the edit options should be linked to the search routines.

What-if. As a subroutine to the search and edit routines, the user should be able to make changes to the database or individual records on a trial basis. These trials, or what-ifs can be done in a separate region of the spreadsheet or, if they are done in the main database, an "undo" feature to reverse the actions must be available.

Compare. The compare menu accesses the search routines to allow records or groups of records to be compared. The basis of comparison will be user-specified. The display of the comparison will be selected from the display menu.

Display. This menu will offer options for displaying the data. It must be linked to the search and the compare routines. Options may include the display of summary data only, the display of raw data in the input region, and various forms of graphic displays.

Data. As the user becomes more sophisticated, statistical calculations on the data may be desired. Because these calculations may be done on individual records to track their history or on groups of records to better define the measurements, this routine must also be linked to the search routines.

This database can be the most useful store of information maintained by the pipeline company. The usefulness will, to a large degree, hinge on the strength of the data programs. As a company begins to focus on risk management, special attention should be given to the data management process.

Typical Pipeline Products

PRODUCT	Boiling Pt (deg F)	Nh	Nf	Nr	RQ points[1]
Benzene	176	2	3	0	8
Butadiene (1,3)	24	2	4	2	10
Butane	31	1	4	0	2[2]
Carbon Monoxide	−314	2	4	0	2
Chlorine		3	0	0	8
Ethane	−128	1	4	0	2
Ethyl Alcohol	173	0	3	0	4
Ethylbenzene	277	2	3	0	4
Ethylene	−155	1	4	2	2
Ethylene Glycol	387	1	1	0	6
Fuel Oil (#1–#6)	304–574	0	2	0	6
Gasoline	100–400	1	3	0	6
Hydrogen	−422	0	4	0	0
Hydrogen Sulfide	−76	3	4	0	6
Isobutane	11	1	4	0	2[2]
Isopentane	82	1	4	0	6
Jet Fuel B		1	3	0	6
Jet Fuel A & A1		0	2	0	6
Kerosene	304–574	0	2	0	6
Methane	−259	1	4	0	2

Table continued on next page

Table continued

PRODUCT	Boiling Pt (deg F)	Nh	Nf	Nr	RQ points[1]
Mineral Oil	680	0	1	0	6
Naphthalene	424	2	2	0	6
Nitrogen		0	0	0	0
Petroleum-Crude		1	3	0	6
Propane	−44	1	4	0	2
Propylene	−53	1	4	1	2
Toluene	231	2	3	0	4
Vinyl Chloride	7	2	4	1	10
Water	212	0	0	0	0

Source: Dow Chemical [9].
[1] *Based upon 1991 CERCLA Reportable Quantities (RQ) and Figure 7-4 with the following:*

RQ(lbs)	Points
none	0
5000	2
1000	4
100	6
10	8
1	10

[2] *When at temperatures higher than the boiling point.*

Leak Rate Determination

Leak Rate Determination

Fluid flow through pipelines is a complex and not completely understood problem. It is the subject of continuing research by engineers, physicists, and more recently, those studying non-linear dynamic systems, popularly called the science of chaos. As with all parts of this risk assessment tool, we are not concerned with exact numerical solutions, only relative quantities.

In general, fluid flow in pipes is assigned to one of two flow regimes, turbulent or laminar. Some make distinctions between rough turbulent and smooth turbulent, and a region termed the transition zone is also recognized. However, in simplest terms, the flow pattern will be characterized by uniform, parallel velocities of fluid particles—laminar flow—or by turbulent eddies and circular patterns of fluid particle velocities—turbulent flow—or by some pattern that is a combination of the two. The flow pattern is dependent upon the fluid average velocity, the fluid kinematic viscosity, the pipe diameter, and the roughness of the inside wall of the pipe.

Several formulas that relate these parameters to fluid density and pressure drop offer approximate solutions for each flow regime. These formulae make a distinction between compressible and non-compressible fluids. Liquids such as crude oil, gasoline, and water are considered to be non-compressible while gases such as methane, nitrogen, and oxygen are considered to be compressible. Highly volatile products such as ethylene, propane, and propylene are generally transported as dense gases—they are compressed in the pipeline until their properties resemble those of a liquid, but will immediately return to a gaseous state upon release of the pressure.

For purposes of this risk assessment, any consistent method of flow calculation can be used. Because the primary intent here is not to perform flow calculations but rather to quickly determine relative leak quantities, some simplifying parameters are in order:

- Release duration is arbitrarily chosen at ten minutes.
- Complete line rupture (guillotine-type failure) is used.[1]
- Operation at MAOP is taken as the initial condition.
- Initial conditions are assumed to continue for ten minutes.
- Depressurization, flow reductions, etc., which occur during the ten minute release scenario, are ignored.
- An arbitrary transition point from liquid to gas will be chosen for flashing fluids.
- Pooling of liquids and vapor generation from those pools is ignored.
- Temperature effects are ignored in the equations but should be considered in choosing the liquid calculation versus the gas calculation. The evaluator should assume the worst case, for example, a butane release on a cold day versus a hot day.
- Pressure due to elevation effects is considered to be a part of MAOP.

Using these simplifying parameters must not mask a worst case scenario. The parameters are selected to reflect conservative, worst case scenarios. The evaluator must affirm that one or more of the above parameters does not actually reflect a less severe scenario.

Gas Flow

For compressible fluids, a calculation for flow through an orifice can be used to approximate the flow rate escaping the pipeline [7].

---

[1] *Reasoning behind selection of this parameter is provided in Chapter 7.*

$$q = YCA \sqrt{\frac{(2g)\ 144\Delta P}{\rho}}$$

where
Y = expansion factor (usually between 0.65 and 0.95)
A = cross sectional area of the pipe (square feet)
C = flow coefficient (usually between 0.9 and 1.2)
g = acceleration of gravity (32.2 ft per sec per sec)
ΔP = change in pressure across the orifice (psi)
ρ = weight density of fluid (lbs per cubic foot)
q = flow rate (cubic feet per sec)

In the case of a discharge of the fluid to atmosphere (or other low pressure environment), Y can be taken at its minimum value, and the weight density of the fluid should be taken at the upstream condition.

Liquid Flow

For incompressible fluids, the equation of flow through an orifice is essentially the same with the exception of the expansion factor, Y, which is not needed for the case of incompressible fluids [7].

$$q = CA \sqrt{\frac{(2g)\ 144\Delta P}{\rho}}$$

where
A = cross sectional area of the pipe (square feet)
C = flow coefficient (usually between 0.9 and 1.2)
g = acceleration of gravity (32.2 ft per sec per sec)
ΔP = change in pressure across the orifice (psi)
ρ = weight density of fluid (lbs per cubic foot)
q = flow rate (cubic feet per sec)

Alternately, other common liquid flow equations such as the Darcy equation may be used to calculate this flow. A consistent approach is the important thing.

Crane Valve [7] should be consulted for a complete discussion of these flow equations.

Flashing Fluids

Fluids that flash, that is, they transform from a liquid to a gaseous state upon release from the pipeline, pose a complicated problem for leak rate calculation. Initially, droplets of liquid, gas, and aerosol mists will be generated in some combination. These may form liquid pools that continue to generate vapors. The vapor generation is dependent upon temperature, soil heat transfer, and atmospheric conditions. It is a non-linear problem that is not readily solvable. Eventually, if the conditions are right, the liquid will all flash or vaporize and the flow will be purely gaseous.

To simplify this problem, an arbitrary scenario is chosen to simulate this complex flow. Three minutes of liquid flow at MAOP is added to seven minutes of gas flow at the product's vapor pressure to arrive at the total release quantity after ten minutes. This conservatively simulates a situation where, upon pipeline rupture, pure liquid is released until the nearby pipeline contents are depressured from the rupture pressure to the product's vapor pressure. Three minutes at the higher pressure—the initial pressure (MAOP)—simulates this. Then, when the nearby pipe contents have reached the product's vapor pressure, any liquid remaining in the line will vaporize. This vapor generation is simulated by seven minutes of gas flow at the vapor pressure of the pipeline contents.

This is, of course, a gross oversimplification of the actual process. For this application, however, the scenario, if applied consistently, should provide results to make adequate distinctions in leak rates between pipelines of different products, sizes, and pressures.

Pipe Wall Thickness Determination

Wall Thickness Determinations

Some equations and design concepts are presented in this section to give the evaluator who is not already familiar with pipeline design methods a feel for some of the commonly used formulae. This section is not intended to replace a design manual or design methodology. Used with the corresponding risk evaluation sections, this appendix can assist the non-engineer in understanding design aspects of the pipeline being examined.

Pipeline wall thicknesses are determined based upon the amount of stress that the pipe must withstand. Design stresses are determined by careful consideration of all loadings to which the pipeline will be subjected. Loadings are not limited to physical weights such as soil and traffic over the line. A typical analysis of anticipated loads for a buried pipeline would include allowances for:

- internal pressure
- surge pressures
- soil loadings (including soil movements)
- traffic loadings

For each of these loadings, failure must be defined and all failure modes must be identified. Failure is often defined as permanent deformation of the pipe. After permanent deformation, the pipe may no longer be suitable for the service intended. Permanent deformation occurs through failure modes such as bending, buckling, crushing, rupture, bulging, and tearing. In engineering terms, these relate to stresses of shear, compression, torsion, and tension. These stresses are further defined by the directions in which they act; axial, radial, circumferential, tangential, hoop, and longitudinal are common terms used to refer to stress direction. Some of these stress direction terms are used interchangeably.

Pipe materials have different properties. Ductility, tensile strength, impact toughness, and a host of other material properties will determine the weakest aspect of the material. If the pipe is considered to be flexible (will deflect at least 2% without excessive stress) the failure mode will likely be different from a rigid pipe. The highest level of stress directed in the pipe material's weakest direction will normally be the critical failure mode. The exception may be buckling which is more dependent on the geometry of the pipe and the forces applied.

Another way to say this is that the critical failure mode for each loading will be the one that fails under the lowest stress level (and, hence, requires the greatest wall thickness to resist the failure). Overall then, the wall thickness will be determined based upon the critical failure mode of the worst case loading scenario.

Internal Loadings

Internal pressure is often the governing design consideration for pressurized pipelines. The magnitude of the internal pressure along with the pipe characteristics determines the magnitude of stress in the pipe wall (due to internal pressure alone) which in turn determines the required wall thickness. This stress (or the associated wall thickness) is calculated using an equation called the Barlow formula:

$$\sigma_{max} = \frac{P_i \times D}{2 \times t}$$

where σ_{max} = maximum stress (psig)
         P_i = internal pressure (psig)
         D = outside diameter (inches)
         t = wall thickness (inches)

This equation specifically calculates the tangential or hoop stress of a thin walled cylinder (Figure C-1). It assumes that the wall thickness is negligible compared to the diameter. Normally the outside diameter is used in the equation (rather than the average diameter) to be slightly more conservative. An exception is concrete pipe, in which the internal diameter is used in the calculation [19]. This allows for concrete's minimal tensile strength. Barlow's formula is not theoretically exact, but yields results within a few percent of actual, depending upon the D/t ratio (higher D/t yields more accurate results, lower yields more conservative results). (See Merritt, p. 21.35 [19].)

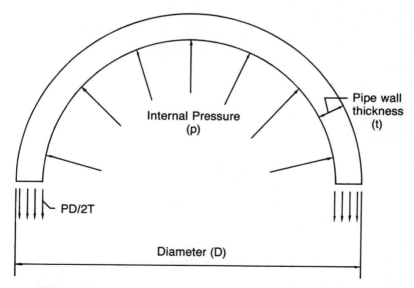

Figure C-1. Barlow's formula for internal pressure stress.

Since many plastic pipe manufacturers refer to a standard dimension ratio (SDR), Barlow's equation can be written using this factor instead of using the diameter and wall thickness separately:

$$SDR = \frac{D_o}{t}$$

$$\sigma_{max} = \left(\frac{P_i}{2}\right) \times (SDR - 1)$$

where SDR = Standard Dimension Ratio
        Do = outside diameter (inches)
          t = pipe wall thickness (inches)
     σ_{max} = maximum stress (psig)
       P_i = internal pressure (psig)

External Loadings

External forces require complex calculations both in determining actual loadings and the pipe responses to those loadings. Soil loads, traffic loadings, and the pipe weight are typical loadings. For offshore and submerged pipelines, the effects of water pressure, currents, floating debris (producing impact loadings), and changing bottom conditions must also be considered. An equation given to calculate required wall thickness to resist buckling due to a static uniform external pressure is [19]:

$$t = D \times \sqrt[3]{\frac{6 \times p}{E}}$$

where t = wall thickness (inches)
      D = diameter (inches)
      p = uniform external pressure (ψ)
      E = pipe modulus of elasticity (ψ)

This equation does not consider the soil-pipe interaction that is a critical part of the buried pipeline system. A rigid pipe must directly withstand the external loads applied. Upon overstressing, typical failure modes are shear and crushing. A flexible pipe, however, deflects under load, allowing the surrounding soil to assist in the support of the load. If this deflection or bending becomes excessive, ring deflection may be the failure mode causing buckling of the flexible pipe. More will be said about this under *Longitudinal Stresses*.

If the external load has a velocity component associated with it, this must also be considered. Highway traffic, rail traffic, and aircraft landings are examples of moving or live loads that, in addition to their static weight, carry an impact factor due to their movement. This impact factor can magnify the static effect of the vehicles' weight.

Design formulae to calculate loadings from moving vehicles can be found in pipeline design manuals.

Longitudinal Stresses

While the primary stress caused by internal pressure is hoop stress, stresses are also produced in other directions. The longitudinal stress produced by internal pressure can be significant in some pipe materials. The amount of restraint on the pipeline in the longitudinal direction will impact the amount of longitudinal stress generated in the pipe. If the pipe is considered to be completely restrained longitudinally, the magnitude of the longitudinal stress is directly proportional to the hoop stress. The proportionality factor is called Poisson's coefficient or ratio. Some values of Poisson's ratio are:

Steel 0.30
Ductile Iron 0.28
PVC 0.45
Aluminum 0.33

If the pipe is considered to be unrestrained longitudinally, the longitudinal stress is numerically equal to about one-half of the hoop stress. In most cases, the actual stress situation is somewhere between the totally restrained and totally unrestrained conditions. A rule of thumb for buried steel pipelines shows that the longitudinal stress generated by internal pressure can be approximated by [30]:

$$S_l = 0.45 \times S_t$$

where S_l = longitudinal stress
          S_t = tangential stress

Longitudinal stresses also occur as a result of differential temperatures. These stresses can be calculated from:

$$\sigma_{temp} = -\alpha \times (\Delta T) \times E$$

where σ_{temp} = temperature induced longitudinal stress
        α = linear coefficient of expansion
      ΔT = temperature change
       E = modulus of elasticity of pipe material

Bending stresses are caused by deflection of the pipe. Inadequate lateral support of the pipeline can therefore allow axial bending and hence longitudinal stress (Figure C-2). Inadequate support can be caused by:

- uneven excavation during initial construction
- undermining due to subsurface water movements
- varying soil conditions that allow the differential settling

In general, flexible pipes are less susceptible to damage from these causes because the pipe can deflect and adjust to changing lateral supports. In the case of either flexible or rigid pipes, design considerations must be given.

Beam formulas are usually used to calculate bending stresses. Assumptions are made as to the end conditions because this is a critical aspect of the beam calculations. Whether or not the pipe is free to move in the longitudinal direction determines how much bending stress

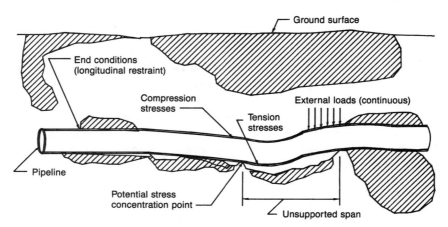

Figure C-2. Bending stresses.

is generated. In the case of buried pipelines, the end condition—the freedom of movement in the longitudinal direction—is dependent upon the amount of pipe-to-soil bonding, and the pipeline configuration (nearby bends or valves may act as anchors to restrict movements).

In general, hoop stresses are independent from longitudinal stresses. This means that the most severe stress will govern—the stresses are not additive. The exception occurs if the longitudinal stress becomes compressive and must then be added to the hoop pressure stress.

A third category of pipe stresses is radial stress. Radial stresses are usually considered to be negligible in comparison with hoop and longitudinal stresses.

Other Considerations

Depending upon the pipe material, other criteria may govern wall thickness calculations. Buckling, cracking, deflection, shear, crushing, vacuum collapse, etc. may ultimately determine the wall thickness requirements. More specific formulas are available for detailed analysis of loadings associated with these failure modes.

In all pipe materials, special allowances must be made for stress risers. Notches, cracks, or any abrupt changes in wall thickness or shape, can amplify the stress level in the pipe wall. See further discussions under Fracture Toughness in the *Design Index.*

APPENDIX D

Surge Pressure Calculations

Surge Pressures

Surge pressures, often called water hammer, are caused when a moving fluid is suddenly brought to a stop. The translation of kinetic (moving) energy to potential energy results in an increase in the internal pressure—the creation of a pressure wave.

The magnitude of the pressure increase is found with the following equation [18]. Surge pressure in feet of water is readily converted to psig by multiplying by 0.43 psig/feet of water.

$$\Delta H = \left(\frac{a}{g}\right) \times \Delta V$$

where ΔH = surge pressure, (feet of water)
$\quad\quad\quad a$ = velocity of the pressure wave (feet/sec)
$\quad\quad\quad g$ = acceleration due to gravity (32 ft/sec^2)
$\quad\quad\quad \Delta V$ = change in velocity of fluid (feet/sec)

We can see from this equation, that the magnitude of the pressure surge is directly related to the speed of the pressure wave and the fluid velocity change.

To calculate the speed of the pressure wave in the pipe, we can use the following equation [19]:

$$a = 12 \times \frac{\sqrt{\dfrac{K}{\rho}}}{\sqrt{1 + \left(\dfrac{K}{E}\right) \times \left(\dfrac{D}{t}\right) \times C_1}}$$

where
a = pressure wave velocity (feet/sec)
K = bulk modulus of the fluid (lb/in^2)
ρ = density of the liquid (slugs/ft^2)
D = internal diameter of pipe (in)
t = pipe wall thickness (in)
E = modulus of elasticity of pipe material (lb/in^2)
C_1 = constant dependent upon pipe constraints

We can see from this equation that pressure wave speed is dependent upon pipe properties (diameter, thickness, modulus of elasticity) as well as fluid properties (bulk modulus, density). This means that the pressure wave will travel at different speeds depending not only upon the product, but also upon the pipeline itself. A more elastic pipe material slows down the pressure wave. As the diameter-to-wall thickness ratio increases, the wave speed decreases.

Because fluid compressibility is dependent upon density and bulk modulus, we can see that the pressure wave speed varies inversely with the compressibility. Fairly incompressible fluids will support faster pressure waves and, hence, greater surge potentials. Note that hydrocarbons are far more compressible than water.

Another component of the pressure surge calculations should be the wave attenuation. Due to friction losses in the pipeline, the pressure wave will be dampened as it travels. This reduction in pressure magnitude with distance travelled can be calculated and becomes a consideration in pipeline design.

The above equations assume instantaneous fluid velocity changes. If the abruptness of the velocity change is controlled, the maximum surge pressure is also controlled. A common example is the rate of closure of a valve. A slamming shut of the valve effectively brings the velocity to zero instantly. A gradual closure causes small, incremental velocity changes with corresponding small surges. How fast is too fast? The following equation allows a critical time to be calculated [19]:

$$T_{cr} = \frac{2 \times L}{a}$$

where T_{cr} = critical time (sec)
         L = distance of pressure wave travel before
                reflection(ft)
         a = velocity of pressure wave (ft/sec)

The critical time is the maximum flow stoppage time that will still allow the maximum surge pressure. Flow stoppage times that are higher than this value produce smaller surge pressures. This critical time is dependent upon the piping configuration because the reflection time from the initiating event governs the calculation.

It is important to note that, in the case of valve closures, the flow stoppage is not necessarily proportional to the actual amount of closure. A gate valve, for instance, may cause 90% of the flow stoppage within the last 10%–50% of the gate travel. The designer must consider the effective closure time as opposed to the actual closure time.

Pressure surges may be caused by valve closures, pumps starting or stopping, the sudden meeting of fluid columns moving at different velocities, and other phenomena that abruptly change the velocity of the pipe fluid. There are many design options to allow for pressure surge effects including relief valves, surge tanks, valve closure controls, pump by-passes, and heavier walled pipe at critical sections.

Pipeline Risk Evaluation Example

The following is an overall example of how this risk evaluation works. A hypothetical pipeline section is being scored according to the suggested weightings provided in this book. Paragraphs describing the pipeline and its environment present information that has been given to the evaluator by the pipeline operator. Following every few paragraphs is a schedule showing how the evaluator has scored the section based on the information provided.

Example

The pipeline section being evaluated is two miles in length, buried approximately 36 in. except for a 200-ft shallow section that is only buried to 30 in. One mile of this section passes through a populated area. There is one aboveground valve station in a remote location. The valve station is clearly marked with signs.

The pipeline operator participates in a highly regarded ULCCA approved one-call system. One-call systems are mandated by this state's law. Pipeline company personnel are assigned to handle one-call reports immediately upon receipt of the report.

There is a public education program in place that includes mailouts and door-to-door annual contact with adjacent residents.

Not all roads are marked with signs indicating the presence of the pipeline. There are some stretches of right-of-way that are overgrown with vegetation. The entire section is patrolled by aircraft once a week.

I. Third Party Index
A.	Minimum Depth of Cover	30/3 =	10 pts A*
B.	Activity Level	High activity	0 pts A
C.	Above-Ground Facilities	0 + 5 + 1 =	6 pts A
D.	On-Call System	4+2+2+2+5=	15 pts P
E.	Public Education	2 + 4 =	6 pts P
F.	Right-of-Way Condition		2 pts P
G.	Patrol Frequency		6 pts P

Total 45 pts

*A = Attribute   Attribute total 16 pts
 P = Prevention   Prevention total 29 pts

This two-mile section contains three cased road crossings. These must be considered to be atmospheric exposures. The only other exposure to the atmosphere in this section, is the ground/air interface exposure. This is where the pipe comes above grade at the valve station. There are no supports—the pipe itself provides the structural support for the valves. The valve station is exposed to hot, humid weather and is close to an industrial complex. The station is frequently re-painted by professional painters.

The natural gas being transported is thought to be mostly dry and free from corrosive impurities. There is, however, no monitoring being done to confirm this. This pipeline section has never been pigged.

A cathodic protection system is in place. System checks including bi-monthly rectifier inspections and anode bed checks are performed to ensure malfunctions are detected. There are test leads at the three casing pipes and at each foreign pipe crossing (there are four in this section). The maximum distance between test leads is .5 mile. Pipe-to-soil potential readings are taken at these leads every three months.

The pipeline is 12 years old. While the coating was most likely properly applied prior to installation, exposing the pipe two years ago revealed some coating damage. This damage might have occurred during original installation or it may have been caused by movements of the expansive clays that predominate in this area. Heavy soil moisture is common, but there is no evidence of damaging microorganism activity.

The pipeline parallels a high power electric line for approximately 1,000 ft. The distance between the pipeline and powerline in this stretch is about 150 ft. The operator has given no consideration to the potential for AC induced corrosion.

A close interval survey was performed five years ago. Corrosion experts interpreted the survey data and concluded that the pipeline in this section was well protected by the cathodic protection currents. No internal inspection has been performed.

II. Corrosion Index = (Atmospheric Corrosion)   20%
    + (Internal Corrosion)   20%
    + (Buried Metal Corrosion)   60%

100%

A. Atmospheric Corrosion
    1. Facilities   1 – 1 (multiple casings)   0 pts A
    2. Atmosphere   hot/humid/chemical   3 pts A
    3. Coating/Inspection   good   5 pts P

8 pts

B. Internal Corrosion
    1. Product Corrosivity   unknown/possible   7 pts A
    2. Internal Protection   none   0 pts P

7 pts

C. Buried Metal Corrosion
    1. Cathodic Protection   8 pts P
    2. Coating Condition   fair   6 pts P
    3. Soil Corrosivity   high   0 pts A
    4. Age of System   10 – 20 years   1 pt A
    5. Other Metals   7 with monitoring   4 pts A
    6. AC Induced Current   0 pts A
    7. Mechanical Corrosion   7 + low stress   4 pts A
    8. Test Leads   3 + 3 =   6 pts P
    9. Close Interval Survey   8 – 5 =   3 pts P
    10. Internal Inspection Tool   0 pts P

32 pts

Total Corrosion Index 8 + 7 + 32 = 47 pts
Attribute Total 19 pts
Prevention Total 28 pts

The MAOP of this pipeline is 1400 psig. This MAOP was confirmed by hydrostatic testing immediately after initial construction (12 years ago). The test pressure was 2100 psig, maintained for 24 hours, and validated by an independent engineer.

The pipeline has 6-inch pipe with 0.25-inch wall thickness (nominal) of Grade B (Specified Minimum Yield Strength (SMYS) = 35,000 psi) pipe. Wall thickness measurements taken several years ago indicate that the actual wall thickness (due to manufacturing tolerances) may be as low as 0.23 inches. All road crossings are encased. No unusual external loadings are seen. Required wall thickness is calculated to be

$$t = [P \times D)/(2 \times SMYS)] + 10\% = 0.146 \text{ inch}$$

The valve station in this section has flanges that are rated for operation at 1480 psig of pressure. All other components, including the pipe itself, are rated for this pressure or higher.

This section normally operates at 900 psig or less. The pressure is usually constant, fluctuating only 100 to 150 psig monthly. Fatigue cycles are therefore limited to about 12 per year at cycle magnitudes of about 150/1400 = 11% of MAOP. Water hammer, or surge potentials 10% MAOP are not possible with the natural gas being transported.

The expansive soils show wide ground cracks when the soil becomes quite dry. Foundation and house slab cracking is common in the area. Except for damages to the pipe coating, however, soil movements are not thought to present a serious threat to this pipeline.

III. Design Index
A.	Pipe Safety Factor	$[(0.23/0.146) - 1] \times 20 =$	11 pts A
B	System Safety Factor	$[(1480/1400) - 1] \times 20 =$	1 pt A
C.	Fatigue	150 cycles at 11% MAOP =	13 pts P
D.	Surge Potential		10 pts P
E.	System Hydrotest	$[(2100/1400) - 1] \times 30 + 0 =$	15 pts P
F.	Pipe Movements	low potential	6 pts A

Total 56 pts
Attribute total 18 pts
Prevention total 38 pts

Although no formal fault tree or event tree analyses have been performed, it is thought to be an extremely remote chance that MAOP can be exceeded. The pressure sources into this line (producing fields and foreign pipelines) could only overpressure the line if gas was packed into the line for many hours. A blocked main line valve and failure of all safety devices with the situation being uncorrected for many hours is the only plausible scenario for overpressure to occur. The valve station in this section of pipeline has an automatic closure device, set to close upon rapid pressure drop, for line break protection. There are no devices to prevent overpressure in this section. Safety devices on the pressure sources are owned by others, are not inspected by this pipeline operator, and are not completely redundant. The operator does monitor pressures and flows upstream and downstream of this section. Monitoring is done through a SCADA system in a control room manned 24 hours per day.

Records from initial design and construction are weak. Some pipeline personnel working today participated in the design and construction inspection. These individuals seem knowledgeable with regards to good design and installation processes and feel that a good level of professionalism was employed. Unfortunately, documentation does not exist to support this. The evaluator believes that the expertise and professionalism were present to ensure a quality pipeline job, but because record-keeping is so poor, he feels that a degree of uncertainty exists and awards 50% of the points possible for the design and construction items. This part of the evaluation will apply to all pipelines operated by this operator, not just this particular pipeline section.

The operator does not have a formal system of job procedures. Most technicians have personal notes from when they were initially trained. These notes serve as guidebooks, but are not standardized and do not replace operating procedures.

Field personnel are in close and constant communications with the control room personnel. All pipeline personnel are informally trained—on the job training—and documentation is sporadic. All personnel do participate in a strong company-wide safety program. Regular emergency drills are a part of this safety program. A drug-testing program is in place for pre-job screening, testing for cause, and random testing of individuals in critical positions.

The pipeline operator performs and documents regular surveys including river crossing profiles, corrosion control, and population surveys.

Critical instruments are clearly marked (painted red) and are maintained per a rigid testing and calibration schedule. Although procedures do not exist, the schedule and documentation associated with this and other maintenance activities is strong. No mechanical devices are present to prevent accidental disabling of an instrument.

IV. Incorrect Operations Index
A. Design P
   1. Hazard Identification   2 pts
   2. MAOP Potential   extremely unlikely   10 pts
   3. Safety Systems $3 + 1 - 3 =$ 1 pt
   4. Material Selection   1 pt
   5. Checks   1 pt
                                                  15 pts

B. Construction P
   1. Inspection
   2. Materials
   3. Joining
   4. Backfill
   5. Handling
   6. Coating
                                                  10 pts

C. Operation P
   1. Procedures   1 pt
   2. SCADA/Communications   5 pts
   3. Drug testing   2 pts
   4. Safety Programs   2 pts
   5. Surveys   2 pts
   6. Training   2 pts
   7. Mechanical Error Preventers   0 pts
                                                  14 pts

D. Maintenance P
   1. Documentation   2 pts
   2. Schedule   3 pts
   3. Procedures   1 pt
                                                  6 pts

Incorrect Operations Index $15 + 10 + 14 + 6 = 45$ pts
Prevention Total 45 pts
Attribute Total 0 pts

Total Index Sum 45 + 47 + 56 + 45 = 193 pts

Total Attribute Sum = 16 + 19 + 18 = 53

Total Prevention Sum = 29 + 28 + 38 + 45 = 140

The natural gas transported in this pipeline is 97% pure methane. The non-methane components will not alter the product hazard calculations. The 10-minute leak volume for the complete line rupture scenario is calculated to be over 100,000 lbs. The molecular weight of methane is 16.

V. Leak Impact Factor
A. Product Hazard (Acute + Chronic Hazards)
1. Acute Hazards
   a. N_f 4
   b. N_r 0
   c. N_h 1

$$\overline{\text{Total } (N_h + N_r + N_f)\ 5}$$

2. Chronic Hazard, RQ 2

$$\text{Product Hazard} = 5 + 2 = 7$$

B. Dispersion Factor (Spill Score) ÷ (Population Score)
1. Vapor Spill   MW = 16, >500,000 lbs   3
2. Population Density   Class 3   3

$$\text{Dispersion Factor} = 3/3 = \overline{1}$$

Leak Impact Factor = (Product Hazard)/(Dispersion Factor)
$$= 7/1 = 7$$

Relative Risk Score = (Index Sum)/(Leak Impact Factor)
$$= 193/7 = 27.6$$

Analysis of the Scores

While the scores calculated above have the most meaning in the context of other pipeline evaluations, the operator can still use these numbers alone. The evaluation can be summarized as follows:

Third Party Index	45 points
Corrosion Index	47 points
Design Index	56 points
Incorrect Operations Index	45 points
Total Index	193 points
Leak Impact Factor	7
Relative Risk Score 170/7 =	27.6

If the operator wants to improve his position—manage the risk to improve reliability and safety—he may wish to start with the lowest scored index. Recall in the evaluation that this particular operator is weak in procedures and training. This model suggests that more emphasis on formalized training and the development and use of procedures will reduce the human error risk. This reduction would be quantified in the *Incorrect Operations Index.*

Similarly, looking at the *Design Index,* a hydrostatic test will increase the index score by 10 points. Clearing the right-of-way and reducing the exposure of the shallow stretch will each increase the *Third Party Index.* Prevention items are the easiest (and sometimes only way) to reduce risk.

A simple economic analysis can be performed to compare the costs of each of these actions with the benefits. The benefits for each are quantified in this model through the point scoring.

The quality of this pipeline section and its operation can be determined only through comparisons. Standards that reflect company policy can be developed for comparison purposes. Similar sections can be evaluated to create the database for comparisons. The attributes vs preventions distinction will ensure that the most meaningful comparisons are made.

Glossary

This glossary defines terms as they are used in this text. In some cases, the definitions may differ slightly from strict dictionary definitions.

Acute Hazard. A potential threat whose consequences occur immediately after initiation of an event. Examples include fire, explosion, and contact toxicity.

Anode. A component of a corrosion cell, the anode is the metal that gives up ions and loses mass during the corrosion process.

Automatic valve (also called automatic block valve). A mechanical device that prevents flow in a pipeline and is designed to operate when it receives a predetermined signal. The signal is transmitted without human action. See also *Remotely Operated Valve*.

Backfill. The soil that is placed over the pipe as one of the final steps in pipeline installation. Sand is often used as a backfill material because of the uniform support it provides and because it does not damage the pipe coating during installation.

Cathode. A component of a corrosion cell, the cathode is the metal that attracts ions and gains mass through the corrosion process.

Cathodic protection. A method of corrosion prevention in which a low voltage charge is impressed on a metal in order to cause the metal to behave as a cathode and, hence, be protected from corrosion.

Chronic Hazard. A potential threat that can continue to cause harm long after the initial event. Examples include carcinogenicity, groundwater contamination, and long-term health effects.

Check valve. A mechanical device that prevents pipeline flow in one direction only. Flow is allowed in the opposing direction.

Coating. A material that is placed around and adheres to a pipeline component to protect that component from contact with a potentially harmful substance.

Corrosion. The wearing away of a material, usually by a chemical reaction.

Dispersion Factor. A number that represents one aspect of the relative severity of a pipeline leak. This number scores the spill characteristics using the size of the leak and the nearby population density. It is used to arrive at the Leak Impact Factor.

DOT. Department of Transportation. The regulatory agency of the U.S. government that is charged with regulating aspects of pipeline design, construction, and operation.

EPA. Environmental Protection Agency. The regulatory agency of the U.S. government that is charged with regulating activities that may be harmful to the environment.

Failure. The point at which a structure is no longer capable of serving its intended purpose. While a pipeline that is actually leaking product is the most obvious indication of failure, failure is often also defined as the point at which the material is stressed beyond its elastic or yield point—it does not return to its original shape.

Fatigue. The process of repeated application and removal of stress. Because fatigue can cause a failure to occur at a relatively low stress level, materials that must resist such cycles of stress must be specially designed for this service.

Fracture toughness. The ability of a material to resist cracking. Materials that are more ductile can absorb larger amounts of energy before cracks spread. Lead has high fracture toughness; glass has low fracture toughness.

Hazard. A potential event that can lead to a loss of life, property, income, etc.

HAZ. Heat Affected Zone. The area of metal around a weld that has been metallurgically altered by the heat of the welding process. This area is often more susceptible to cracking than the parent metal.

Index. One of four general categories to which pipeline accidents can be attributed. Aspects of pipeline design, operation, and environment are scored to arrive at numerical values for the Third Party Index, Corrosion Index, Design Index, and Incorrect Operations Index.

Internal corrosion. Any form of corrosion that occurs on the inside wall of the pipe or internal surfaces of any pipeline component.

Leak Impact Factor. A number that represents the overall consequence of a pipeline failure in the risk assessment methodology presented in this book. This factor is a score based upon the product hazard and the dispersion factor. The Leak Impact Factor is divided into the sum of the four index values to arrive at the relative risk score.

MAOP. Maximum Allowable Operating Pressure (also called MAWP for Maximum Allowable Working Pressure). The highest internal pressure to which the pipeline may be subjected based upon engineering calculations, proven material properties, and governing regulations.

PSI (PSIG and PSIA). Pounds per Square Inch (Gauge or Absolute). This is the normal unit of pressure measurement in the U.S. PSIG is the gauge pressure and is the reading that is seen on a pressure gauge calibrated to zero under atmospheric pressure. PSIG therefore does not separate atmospheric pressure from the reading seen on the gauge. Zero psig is equal to about 14.7 psia, depending upon the exact atmospheric pressure of the area.

Pig. A device designed to move through a pipeline for purposes of cleaning, product separation, or information gathering. A pig is usually propelled by gas or liquid pressure behind the pig. The name "pig" is said to have originated from the sound the device makes as it moves through the pipeline.

Pressure Relief Valve. Also called a "pop valve" or a "safety valve," this class of mechanical safety device is designed to operate at a predetermined pressure to reduce the internal pressure of a vessel. The valve is often designed to close again when the vessel pressure is again below the set point.

Product Hazard. A numerical score that reflects the relative danger of the material being transported through the pipeline. This relative ranking of the product characteristics considers acute and chronic hazards such as flammability, toxicity, and carcinogenicity.

Public Education. The program sponsored by pipeline companies to teach the general public about the pipeline industry. The emphasis is usually on how to avoid and report threats to the pipeline and what precautions to take should a leak be observed.

Relative Risk Value (Rating or Score). The final output of the risk evaluation process presented in this book. This number represents the relative risk of a section of pipeline in the environment and operating climate considered during the evaluation. This score is meaningful only in the context of other scores from pipelines evaluated by this same process.

Rectifier. A device that converts AC electricity into DC electricity and delivers the current onto the pipeline for purposes of cathodic protection.

Release Quantity. This is the quantity of spilled material that will trigger an EPA investigation. Possible categories are 1, 10, 100, 1000, and 5,000 pound spills. More hazardous substances trigger at lower release amounts. For this risk assessment model, release quantities have been assigned to substances not normally regulated by EPA.

Remotely Operated Valve. A mechanical device that prevents flow in a pipeline and is designed to operate upon receipt of a signal transmitted from another location.

Risk. The probability and consequences of a hazard.

ROW. Right of Way. The land above the buried pipeline (or below the aboveground pipeline) that is under the control of the pipeline owner. This is usually a strip of land several yards wide that has been leased or purchased by the pipeline company.

Safety device. A pneumatic, mechanical, or electrical device that is designed to prevent a hazard from occurring or to reduce the consequences of the hazard. Examples include pressure relief valves, pressure switches, automatic valves, and all automatic pump shutdowns.

SCADA. Supervisory Control and Data Acquisition. A system to gather information such as pressures and flows from remote field locations and regularly transmit this information to a central facility where the data can be monitored and analyzed. Through this same system, the central facility can often issue commands to the remote sites for actions such as opening and closing valves and starting and stopping pumps.

SCC. Stress Corrosion Cracking. This is a potential failure mechanism that is a combination of mechanical loadings (stress) and corrosion. It is often an initiating or contributing factor in fatigue failures.

SMYS. Specified Minimum Yield Strength. The amount of stress a material can withstand before permanent deformation (yielding) occurs. This value is obtained from the manufacturer of the material.

Stress. The internal forces acting upon the smallest unit of a material, normally expressed in psi (in the U.S.). When an external loading such as a heavy weight is placed on a material, a level of stress is created in the material as it resists deformation from the load.

Surge pressure. Also referred to as "water hammer." This is a phenomenon in pipeline operations characterized by a sudden increase in internal pressure. This surge is often caused by the transformation of kinetic energy to potential energy as a stream of fluid is suddenly stopped.

Wall Thickness. The dimension measurement between a point on the inside surface of the pipe and the closest point on the outside surface of the pipe. This is the thickness of the pipe material.

Yield Point. In general, this is the point, defined in terms of an amount of stress, at which inelastic deformation takes place. Up to this point, the material will return to its original shape when the stress is removed; past this point, the stress has permanently deformed the material.

References

1. Baker, W. E., et al. *Explosion Hazards and Evaluation,* New York: Elsevier Scientific Publishing Company, 1986.
2. Battelle Columbus Division. *Guidelines for Hazard Evaluation Procedures,* New York: American Institute of Chemical Engineers, 1985.
3. Bolt, Rein and Logtenberg, Theo. "Pipelines Once Buried Never to be Forgotten," *Reliability on the Move: Safety and Reliability in Transportation,* ed. G. B. Guy, London: Elsevier Applied Science, 1989, pp. 195–207.
4. Briggum, S., Goldman, G. S., Squire, D. H., Weinberg, D. B. *Hazardous Waste Regulation Handbook,* New York: Executive Enterprises Publications Co., Inc., 1985.
5. Caldwell, Joseph C. "Pipe Line Safety Arena," *Pipe Line Industry,* November 1990, p. 15.
6. Clarke, N.W.B. *Buried Pipelines,* London: Maclaren and Sons, 1968.
7. Crane Valve Company. *Flow of Fluids Through Valves, Fittings, and Pipe,* Crane Technical Paper No. 410. New York, 1986.
8. *DIN 2413.* Deutsche Normen, Berlin, June 1972.
9. Dow Chemical. *Fire and Explosion Index Hazard Classification Guide,* sixth edition, Dow Chemical Co., May 1987.
10. Dragun, James. *The Soil Chemistry of Hazardous Materials,* Silver Spring, Maryland: Hazardous Materials Control Research Institute, 1988.
11. Esparza, E. D., et al. *Pipeline Response to Buried Explosive Detonations, Volumes I and II,* American Gas Association,

Pipeline Research Committee Final Report AGA Project PR-15-109, Southwest Research Institute Final Report SWRI project 02-5567, August 1981.

12. Federal Register. *Rules and Regulations,* Vol. 54, No. 155, August 14, 1989, pp. 33420–33424, August 30, 1989, pp. 35989–90.

13. Flinn, R. A. and Trojan, P. K. *Engineering Materials and Their Applications,* 3rd edition, Boston: Houghton Mifflin Company, 1986, pp. 513–560.

14. Gleick, James. *Chaos,* New York: Penguin Books, 1988.

15. Hanna, S. R. and Drivas, P. J. *Guidelines for Use of Vapor Cloud Dispersion Models,* New York: American Institute of Chemical Engineers, 1987.

16. Keyser, C. A. *Materials Science in Engineering,* 3rd edition, Columbus: Charles E. Merrill Publishing Company, 1980, pp. 75–101, 131–159.

17. Larsen, K., et. al. "Mitigating Measures for Lines Buried in Unstable Slopes," *Pipe Line Industry,* October 1987, pp. 22–25.

18. Megill, R. E. *An Introduction to Risk Analysis,* 2nd edition, Tulsa: PennWell Books, 1984.

19. Merritt, F. S. *Standard Handbook For Civil Engineers,* New York: McGraw-Hill Book Co, 1976, section 21.

20. NACE. *Recommended Practice: Mitigation of Alternating Current and Lightning Effects on Metallic Structures and Corrosion Control Systems.* National Association of Corrosion Engineers, Nace Standard RP-01-77 (1983 Revision), Item No. 53039.

21. "One-Call Systems," *Pipeline Digest,* March 1991, p. 15.

22. Pipeline Industries Guild. *Pipelines: Design, Construction, and Operation.* London, New York: Construction Press, Inc., 1984.

23. Prugh, R. W. and Johnson, R. W. *Guidelines for Vapor Release Mitigation,* New York: American Institute of Chemical Engineers, 1988.

24. Riordan, M. A. "The IR Drop Paradigm Calls for a Change," *Pipe Line Industry,* March 1991, p. 31–32.

25. Rusin, M., Savvides-Gellerson, Evi. *The Safety of Interstate Liquid Pipelines: An Evaluation of Present Levels and Proposals for Change,* American Petroleum Institute, Research Study 040, July 1987, Washington, DC.

26. Siegfried, Charles "Multiple Uses of ROW for Pipelines," presented at American Gas Association Transmission Conference, May 18, 1971.

27. Smart, Dr. J. S., Smith, G. L. "Pigging and Chemical Treatment Pipelines," Paper presented at *Pipeline Pigging and Inspection Technology Conference,* Feb. 4–7, 1991, Houston, TX.

28. Tuler, S., et. al. "Human Reliability and Risk Management in the Transportation of Spent Nuclear Fuel," *Reliability on the Move: Safety and Reliability in Transportation,* ed. G.B. Guy, London: Elsevier Applied Science, 1989, pp. 167–193.

29. U.S. Dept. of Transportation, Research and Special Programs Administration, Office of Pipeline Safety. *Annual Report of Pipeline Safety—Calendar Year 1988,* 400 Seventh St., S.W. Washington D.C. 20590.

30. Vincent-Genod, J. *Fundamentals of Pipeline Engineering,* Paris: Gulf Publishing Company, 1984.

31. Wheeler, Donald J., Lyday, Richard W. *Evaluating the Measurement Process,* 2nd edition, Knoxville, TN: SPC Press, Inc., 1989.

32. Williams, Peter J. *Pipelines and Permafrost; Physical Geography and Development in the Circumpolar North,* USA: Longman, Inc., 1979.

Index

Boldface page numbers indicate definitions of terms.